Image Processing with Cellular Topology

Vladimir Kovalevsky

Image Processing
with Cellular Topology

 Springer

Vladimir Kovalevsky
Computer Science Department
University of Applied Sciences Berlin
Berlin, Germany

ISBN 978-981-16-5774-0   ISBN 978-981-16-5772-6   (eBook)
https://doi.org/10.1007/978-981-16-5772-6

This Springer imprint is published by the registered company Springer Nature Singapore Pte Ltd.
The registered company address is: 152 Beach Road, #21-01/04 Gateway East, Singapore 189721,
Singapore

Abstract

The book explains why the definition of the boundary by means of the 4- and 8-neighborhood is wrong and suggests the use of the classical topological definition of the boundary while the digital image should be considered as an abstract cell complex. This approach has great significance in digital image processing. It allows a topological justification of many terms used in image processing. However, what is more important from the practical point of view is that we show how to work with cell complexes without the need of a large additional memory. We also suggest a graphical representation of boundaries in a cell complex. Three algorithms for tracing and encoding boundaries in binary, indexed, or color images are described. The code is free of loss of information so that the image can be exactly reconstructed from the code. The book also describes the theory of digital straight segments and an algorithm for dissolving a digital curve in digital straight segments. Another approach for dissolving digital curves into line segment is the polygonal approximation which is also presented in the book. The book considers different approaches to the detection of edges and suggests a new efficient method of edge detection usable for two- and three-dimensional images. Methods of efficiently encoding edges are suggested. Also, boundaries of subsets in a 3D space called surfaces are considered. Algorithms for efficiently encoding surfaces and for reconstructing 3D sets from the codes of all surfaces contained in this set are suggested. In the last chapter, the author suggests discussing the use of the classical definition of the derivative as the limit of the relation of the increment of the function divided by the increment of the argument while the latter tends to zero. This definition cannot be used for estimations of derivatives of non-analytical functions because it becomes wrong at small increments of the argument. Suggested is a useful method using an optimal value of the increment.

Contents

About the Author

Vladimir A. Kovalevsky received his diploma in physics from the Kharkov University (Ukraine) in 1950, the first doctoral degree in technical sciences from the Central Institute of Metrology (Leningrad) in 1957, and the second doctoral degree in computer science from the Institute of Cybernetics of the Academy of Sciences of Ukraine (Kiev) in 1968. From 1961 to 1983 he served as Head of Department of Pattern Recognition at that Institute.

He has been living in Germany since 1983. From 1983 to 1989 he was researcher at the Central Institute of Cybernetics of the Academy of Sciences of the GDR, Berlin. From 1989 to 2004 he was professor of computer science at the University of Applied Sciences Berlin with an interruption for three years (1998–2001). In that time, he was scientific collaborator at the University of Rostock. He worked as visiting researcher at the University of Pennsylvania (1990), as visiting professor of computer sciences at the Manukau Institute of Technology, New Zealand (2005), at the Department of Automatic Control of the "Centro de Investigación y de Estudios Avanzados del IPN" in Mexico (2006), at the University of Applied Sciences Wildau, Germany (2008), and at the Chonbuk National University, Korea (2009). He has been plenary speaker at conferences in Europe, America, and New Zealand. His research interests include digital geometry, digital topology, computer vision, image processing, and pattern recognition. He has published four monographs and more than 180 journal and conference papers in image analysis, digital geometry, and digital topology.

Chapter 1
Introduction

Abstract This chapter explains that many notions used in digital image processing, like connectedness, boundary etc. are topological notions. Attempts to define these notions without using topology were not successful. It is necessary to define a topological space assigned to the image to consistently define these notions. This space must be a finite one to be explicitly representable in the computer. It is explained how such a finite topological space can be constructed and assigned to a digital image.

Keywords Boundary · Topology · Pixel adjacency · Connectivity paradox · Finite topological space · Boundary definition · Separation axiom · Abstract cell complex

We need in image analysis and in computer graphics the notions of connected subsets, of adjacent subsets, of boundaries, of the interior and exterior etc. All these are topological notions important for image processing. Topology may be considered as the geometry of a rubber sheet: all mentioned properties of subsets drawn on a rubber sheet remain preserved when one pulls the sheet without tearing it. A mathematician would say that topological properties are invariant under continuous transformations of the space.

The general topology considers spaces in which even the smallest neighborhood of a point contains infinitely many other points. It is obvious that such a space (and even the smallest part of it) cannot be explicitly represented in a computer. Therefore, we need the topology of the so called *locally finite spaces* or Alexandroff spaces [1, 2] whose elements have neighborhoods containing a finite number of elements.

In the 1970th A. Rosenfeld [3] has suggested considering the 4- and the 8-adjacency of pixels in digital images (Fig. 1.1).

He defined an m − path as a sequence of m − adjacent pixels, m being 4 or 8. He also introduced the m-connectivity: a subset S of pixels is m-connected if it contains for any two pixels of S an m-path from one of the pixels to another (Fig. 1.2).

When considering four pixels having all a common corner one may see the well-known *connectivity paradox*: under the 4 − adjacency each of the two "diagonal"

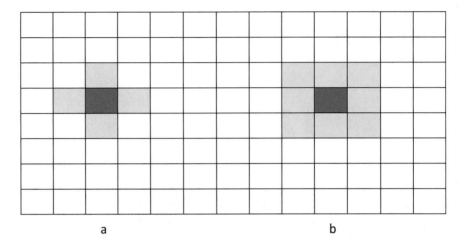

Fig. 1.1 The 4-adjacency (**a**) and the 8-adjacency (**b**)

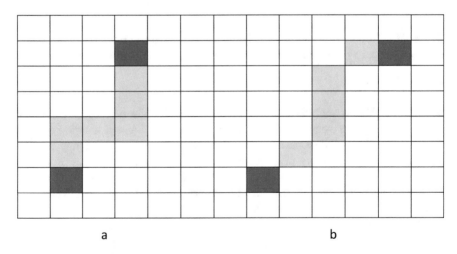

Fig. 1.2 The 4-path (**a**) and the 8-path (**b**) from one dark pixel to another

pairs of pixels is disconnected; under the 8 − adjacency they are both connected. Thus one 8 − path crosses another such path while the two paths have no common elements. Rosenfeld has suggested considering a "mixed" adjacency: the 8 − adjacency for the foreground of an image and the 4 − adjacency for the background or vice versa (Fig. 1.3).

This suggestion has removed the connectivity paradox for binary images. However, it remains for multilevel images. Besides that, a mixed (m, n) − adjacency does not define a topology of the whole space but rather a "topology" of concrete subsets, namely foreground and background of the space: it changes if the subsets change.

Fig. 1.3 The connectivity
paradox: are the dark or the
light pixels connected?

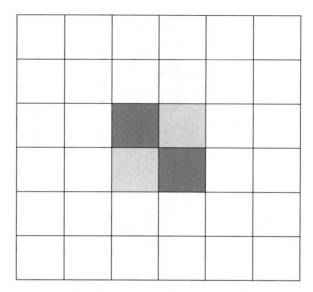

Fig. 1.4 Boundary in the
classical topology: pixel
p belongs to the boundary of
the set *T* because its
neighborhood intersects
both *T* and its complement
S − *T*

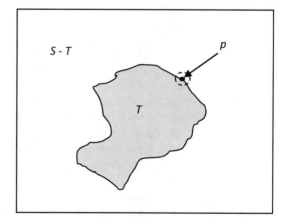

More important for the practical image processing and computer graphics is the
problem of boundaries. *In the classical topology the topological boundary
(or frontier) of a subset T of the space S is the set of all space elements whose
each neighborhood contains both elements of T and of its complement S − T.* In the
topology of continuous spaces, the boundary of a subset of the plane is a curve. It is
thin, i.e., its area is equal to zero. The boundary is the same for *T* and for its
complement *S* − *T* since the above definition is symmetric with respect to *T* and
S − *T* (Fig. 1.4).

When we try to transfer this definition to the case of a digital image considering
the neighborhood of a pixel *P* as the set of all pixels *m*-adjacent to *P* (including
P itself), then the boundary becomes a thick stripe whose width is equal to two pixels
(Fig. 1.5a). When, however, we change the definition so that the boundary of

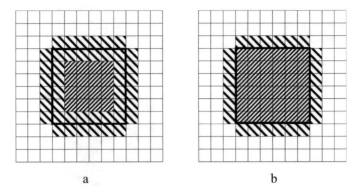

a   b

Fig. 1.5 Boundary defined as a set of pixels whose 4-neighborhood contains pixel both from the foreground and from the background (**a**) and as a set of background pixels whose 4-neighborhood contains pixels from the foreground (**b**)

T contains only pixels of T, then we arrive at two different boundaries: the boundary of T is different from the boundary of the complement $S - T$ (Fig. 1.5b). This is also a topological paradox.

These definitions of a boundary lead to great practical problems. Thus, the excellent modern textbook on image processing by Burger and Burge [4] shows a great problem with tracing boundaries of regions. They describe in Sect. 10 an overly complicated algorithm of simultaneously labeling a connected region and tracing its boundary.

What is the correct solution of the problem?

Imagine, the floor in your bathroom is covered with gray tiles. In the middle, however, there is an area with blue tiles. You want to mark the boundary of the blue area with white color. Would you paint the blue or the gray tiles? Of course, you would paint neither the blue nor the gray tiles, but rather the thin lines between the blue and the gray tiles.

We have the same situation in a digital image: If we want marking the boundary of a subset, we should mark neither some pixels of the subset nor ones of its complement. We must mark the thin lines between the pixels of the subset and those of its complement. These lines belong to the boundary of the set of blue tiles.

Boundary is a topological notion [5]. A point P of the space R belongs to the boundary (or frontier, a topological term for boundary) of a subset S if each neighborhood of P intersects both the subset S and its complement R–S. Therefore, to exactly define a boundary in a digital image we need a topological space containing the pixels of the digital image as space elements.

It is possible, for example, to take a two-dimensional Euclidean space of the size image.width*image.height as such a topological space with real coordinates $x \in [0,$ image.width] and $y \in [0,$ image.height] and integer values ix and iy with $ix \in [0,$ image.width] and $iy \in [0,$ image.height]. We define the set of points (x, y) with coordinates $x \in (ix, ix + 1)$ and $y \in (iy, iy + 1)$ as the two-dimensional cell (2-cell) corresponding to the pixel (ix, iy) of the digital image. This set is a square with

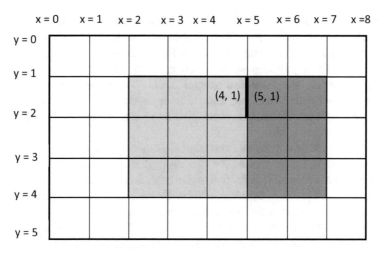

Fig. 1.6 A Euclidean space with the subset being a vertical one-dimensional cell

corners at integer points (ix, iy), $(ix + 1, iy)$, $(ix + 1, iy + 1)$, and $(ix, iy + 1)$, (compare Fig. 1.6). We assign the coordinates (ix, iy) to this cell.

A point with coordinates (ix, y) lies obviously between the 2-cells $(ix - 1, int(y))$ and $(ix, int(y))$. If the pixels $(ix - 1, int(y))$ and $(ix, int(y))$ of the digital image have different colors, then the point (ix, y) belongs to the boundary of the subset of pixels with the color of pixel $(ix - 1, int(y))$ and of the subset of pixels with the color of $(ix, int(y))$. All points with coordinates (x, y) with $x = ix$ and $y \in (int(y), int(y) + 1)$ lie in this boundary. We denote the subset of these points as a vertical one-dimensional cell (1-cell) with coordinates $(ix, int(y))$. This subset is a vertical straight-line segment presented in Fig. 1.6 as a bold segment. The 2-cells $(ix - 1, int(y))$ and $(ix, int(y))$ are adjacent to the 1-cell $(ix, int(y))$. Thus, we say the 1-cell $(ix, int(y))$ bounds the 2-cells $(ix - 1, int(y))$ and $(ix, int(y))$ and is incident with them.

Thus, we denote the subset of points (x, y) with $x = ix$ and $y \in (iy, iy + 1)$ as a vertical one-dimensional cell (1-cell) with coordinates (ix, iy) and the subset of points (x, y) with $x \in (ix, ix + 1)$ and $y = iy$ as a horizontal one-dimensional cell (1-cell) with coordinates (ix, iy). A one-dimensional cell bounds the two adjacent 2-cells and is incident with them.

A point with coordinates (ix, iy) is denoted as a zero-dimensional cell (0-cell) with coordinates (ix, iy). It belongs to the boundary of the subset of pixels with the color C if some of its neighborhoods intersects 2-cells corresponding to pixels with color C and 2-cells corresponding to pixels with some other color. The 0-cell (ix, iy) bounds four adjacent 1-cells with coordinates $(ix - 1, iy)$, $(ix, iy - 1)$ and (ix, iy) and the four adjacent 2-cells with coordinates $(ix - 1, iy - 1)$, $(ix, iy - 1)$, $(ix - 1, iy)$ and (ix, iy). It is incident with the eight called cells.

The set of such defined cells of dimensions from 0 to 2 is a CW complex [6]. If we "forget" the relations of the cells of the CW complex to the Euclidean space, then we obtain an *abstract cell complex*. We retain the bounding and the incident relations.

This transformation is important because it is easier to work with integer coordinates than with real ones while the cells of an abstract complex have integer coordinates and no relations to the real coordinates of a Euclidean space.

It is possible to place an abstract cell complex over the image. A complex contains, besides the two-dimensional cells which are the pixels, also one-dimensional cells which are short thin lines between the pixels. The zero-dimensional cells are the corners of the pixels and simultaneously the endpoints of the one-dimensional cells.

An abstract cell complex is a topological Kholmogorov space [5] and has the topological properties important for our case. Both definitions of the boundary transferring the topological definition and considering the neighborhood of a pixel P as the set of all pixels m-adjacent to P (including P itself), cause problems while the classical topological definition of the boundary is free of problems. The difference between the classical definition and that with an m-adjacency is that a Kholmogorov topological space satisfies the so-called *separation axiom*, which asserts that for any two elements x and y of the space there exists at least one open set and thus at least one neighborhood containing only one of these two elements. This property is obviously not met by neighborhoods defined by means of m-neighborhoods of pixels because if a pixel x belongs to the neighborhood of the pixel y than the pixel y also belongs to the neighborhood of the pixel x. Thus, the abstract cell complex possesses topological properties important for the consistent definition of a boundary.

References

1. Alexandroff P. Discrete spaces. Mat Sbornik. 1937;2:501–18.
2. Alexandroff P, Hopf H. Chapter 3: Topologie der Komplexe. In: Topologie II. Chelsea: University of Kaiserslautern; 1935.
3. Rosenfeld A. Connectivity in digital pictures. J ACM. 1970;17:146–60.
4. Burger W, Burge MJ. Digital image processing. An algorithmic introduction using Java. London: Springer; 2016.
5. Aleksandrov PS. [1956], "Chapter XVIII Topology". In: Aleksandrov AD, Kolmogorov AN, Lavrent'ev MA, editors. Mathematics / its content, methods and meaning. 2nd ed. Cambridge: The M.I.T. Press; 1969.
6. Whitehead JHC. Combinatorial homotopy. I. Bull Am Math Soc. 1949;55(5):213–45.

Chapter 2
Boundary Presentation Using Abstract Cell Complexes

Abstract This chapter describes the properties of abstract cell complexes. The definition of the dimension of cells is explained. It has been shown how the connectivity paradox can be solved by means of complexes. The notion of the smallest open neighborhoods and that of closures are introduced. It is shown how coordinates of cells can be defined, and the difference between the standard and combinatorial coordinates is being explained.

Keywords Abstract cell complex · Bounding relation · Dimension · Connectedness · Solution of the connectivity paradox · Cracks · Points · Smallest neighborhood · Closure · Coordinates of cells · Cartesian complex · Standard and combinatorial coordinates · Frontier · Graphical presentation of boundaries

The problem of consistently defining boundaries can be solved if one considers along a digital image also a corresponding two-dimensional abstract cell complex which is a structure containing besides the pixels which are regarded as two-dimensional cells having the shape of small squares also the one-dimensional cells being the sides of the squares presenting the pixels and the zero-dimensional cells being the corners of the pixels and simultaneously the end points of the one-dimensional cells. Using cell complexes solves the problem because neighborhoods of cells of different dimensions are different; and, if a cell x belongs to the neighborhood of the cell y, then y does not belong to the neighborhood of x as explained below. An abstract cell complex is a topological space and possesses the positive properties of a topological space. It is important that an abstract cell complex satisfies the separation axiom.

2.1   Abstract Cell Complexes

An abstract cell complex is an abstract set of elements called cells provided with an asymmetric, irreflexive and transitive binary relation called the *bounding relation* among its elements. Classical definitions of an abstract cell complex, e.g [1], mention that a non-negative integer number called *dimension* should be assigned to each cell. However, we have found that the dimension can be defined by means of the bounding relation. The dimension of a cell C of an abstract cell complex is equal to the length (number of cells minus 1) of the maximum bounding path leading from any cell of the complex to the cell C. The bounding path is a sequence of cells in which each cell bounds the next one. Figure 2.1 illustrates the definition of the dimension.

Figure 2.1 shows a complex and the bounding relations of its cells. The cell p has dimension equal to 0, or $\dim(p) = 0$, since there is no bounding path leading to the cell p. The path (p, f) leads to the cell f. However, it is not the longest path leading to the cell f: the path (p, e, f) is longer. Its length is 2. Therefore, the dimension of the cell f is 2, or $\dim(f) = 2$. One of the longest bounding paths from p to v is (p, e, f, v). Its length is 3, therefore $\dim(v) = 3$.

The dimension of a complex is the maximum dimension of its cells.

The notion of the bounding path leads to the definitions of the incidence and that of connectedness. If the cell a bounds the cell b, then these cells are incident to each other: the cell a is incident to the cell b and the cell b is incident to the cell a. The incidence relation is symmetric. According to this definition, a bounding path is simultaneously an incidence path: two adjacent cells in a bounding path are incident to each other.

Now we can define the next important notion in the theory of cell complexes: that of connectedness. A subset T of a cell complex C is called connected if and only if for any two elements of T an incidence path exists containing these two elements and

Fig. 2.1 A complex with bounding relations represented by arrows. An arrow points from a to b if a bounds b

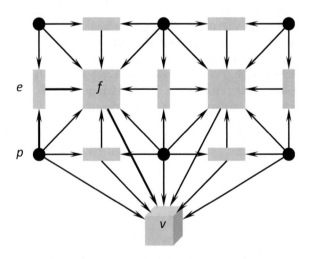

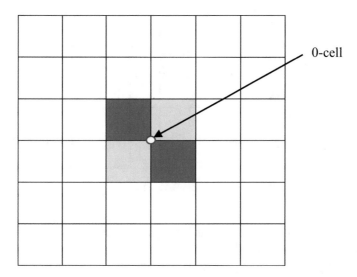

0-cell

Fig. 2.2 Connectivity of the light- and dark-gray pixels depends on the membership of the 0-cell

Fig. 2.3 Bounding relation: the 0-cell a bounds the 1-cell b and the 2-cell c. The 1-cell b bounds the 2-cell c but not the 0-cell a

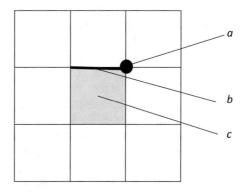

this incidence path lies completely in T. This definition solves the well-known connectivity paradox (see Fig. 1.3 above) which remained unsolved without the theory of abstract cell complexes.

The solution consists in the following: Consider the Fig. 2.2.

If the 0-cell belongs to the same subset of the complex as the light-gray pixels, then the light-gray subset becomes connected and the dark-gray subset disconnected. If the 0-cell belongs to the dark-gray subset, then this subset is connected. If, however, the 0-cell belongs neither to the light-gray nor to the dark-gray subset, then both subsets are disconnected.

A cell can be bounded by another cell of lower dimension. E.g., a cell of dimension 2 (cell c in Fig. 2.3 above) can be bounded by cells of dimension 1 (cell b in Fig. 2.3) and by cells of dimension 0 (e.g., cell a in Fig. 2.3). If the

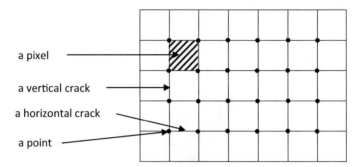

a pixel

a vertical crack

a horizontal crack

a point

Fig. 2.4 Pixels, cracks, and points of a two-dimensional ACC

cell a bounds the cell b, then b cannot bound the cell a. This is the asymmetry property of the bounding relation.

An important notion in the theory of abstract complexes is that of the smallest neighborhood of a cell.

A two-dimensional abstract cell complex (ACC) contains elements of dimensions 0, 1, and 2. When considering an ACC corresponding to a two-dimensional digital image, then the 2-dimensional elements of the ACC correspond to the pixels considered as small squares, its 1-dimensional elements called *cracks* correspond to the sides of these squares, and its 0- dimensional elements called *points* correspond to the corners of the squares (Fig. 2.4). We shall call the 2- dimensional elements of the ACC *pixels* (if it does not lead to a misunderstanding).

Pixels must be 2-dimensional, because a pixel contains color or a gray level being proportional to its area, while the area of a 1- or 0-dimensional element is zero and therefore cannot contain a color or a gray level.

The complex is called "abstract" since its elements, which are called "cells", are not subsets of a Hausdorff space [2] as it is in the case of Euclidean and CW complexes [3]. Cells of an ACC are no subsets of some other space. They are abstract elements.

Now we consider the problem of consistently defining the topological notion of a boundary which is often called in topological literature the *frontier*. The solution of this problem and simultaneously of the problem of consistently defining the connectivity consists in defining *antisymmetric* neighborhoods of the elements, what means that for any two elements x and y of the space there must be a neighborhood containing only one of these two elements: e.g., the neighborhood of x contains x but not y while the neighborhood of y contains both x and y (if x and y are incident). Thus, the elements x and y of different dimensions have different neighborhoods and thus they have different properties. Our ACC corresponding to the digital image must contain elements with *different properties*, while all pixels have the same properties.

This means that our ACC corresponding to the digital image must contain elements of different dimensions possessing such neighborhoods that for any two incident elements x and y the smallest neighborhood of one of these two elements

Fig. 2.5 Smallest
neighborhoods of n-cells c^n
in 2D and 3D complexes

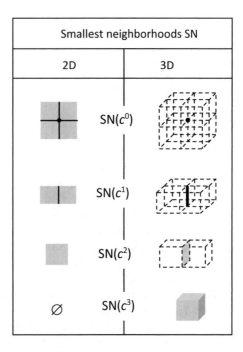

does not contain the other element. For example, the smallest neighborhood of
x does not contain y, if y bounds x.

The smallest neighborhood SN of a cell c of an ACC is the subset containing
c and all cells of the ACC bounded by the cell c (see Fig. 2.5 above).

The smallest neighborhood of a pixel in a two-dimensional ACC is the pixel
itself. Thus, the smallest neighborhood of a pixel consists of a single element and
thus cannot contain other elements.

The smallest neighborhood of a crack in a two-dimensional ACC (not lying at the
border of the image) consists of the crack itself and of two pixels while the crack is
the common side of these two pixels. Thus, one can see that the smallest neighborhood of a crack contains the crack and each of the two pixels, while the smallest
neighborhood of a pixel does not contain the crack.

The smallest neighborhood of a point P in a two-dimensional ACC (not lying at
the border of the image) consists of the point P itself, of four cracks whose one end
point is P, and four pixels whose one corner is P (Fig. 2.5 above). It is easily seen
that this neighborhood also satisfies the separation condition: The smallest neighborhood of a point contains cracks and pixels while the smallest neighborhood of a
crack contains no points. The same is true for the smallest neighborhood of a pixel.

A three-dimensional digital image should be accompanied by a three-dimensional
ACC. It contains besides points, cracks, and pixels the three-dimensional elements
called *voxels*. The smallest neighborhoods of cells of different dimensions in 2D and
3D complexes are shown in Fig. 2.5 above.

Fig. 2.6 Examples (*b*) and (*c*) of closures of the cell *a* depending on the including complex (explanation in text)

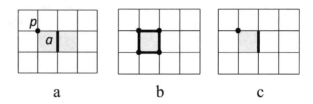

a   b   c

The next important notion is that of the closure. We define first the closure of a set of cells. Let *t* and *T* be subsets of the space *S* such that $t \subseteq T \subseteq S$. Then the set containing together with each cell $a \in t$ also all cells of *T* which bound *a* is called the *closure of t in T* and is denoted by $Cl(t, T)$.

We shall employ the notion of the closure not only for subsets, but also for single cells (singletons). In such cases the notation $Cl(\{c\}, T)$ would be the correct one. However, it is common to employ the notation $Cl(t, T)$ also for singletons to make the notations simpler. It is always clear from the context what is meant.

The closure of a cell *c* of the space *S* is the union of *c* with all cells of *S* bounding c. It is denoted by $Cl(c, S)$.

If the set *S* in the expression $Cl(c, S)$ is known from the context, then it is allowed to drop *S* and simply speak of the closure of a cell *c* and denote it by $Cl(c)$.

The size and shape of the smallest neighborhood and of the closure of a cell depend on the including complex. Consider the following examples in Fig. 2.6.

Figure 2.6a shows a complex *C* represented by squares (shaded and white), thin lines, and vertices of the squares and its subcomplex $S \subset C$ represented by two shaded squares one of which is the 2-cell *a*, one bold line (1-cell) and a small dark disk *p* (a 0-cell). Figure 2.6b shows the closure $Cl(a, C)$ of the 2-cell *a* in the complex *C* while Fig. 2.6c shows $Cl(a, S)$ which is the closure of the 2-cell *a* in the subset *S*. It does not contain the non-bold 1-cells incident to *a* and the three 0-cells not labeled by a small disk since they are not contained in the subset S.

2.2 Coordinates of Cells

We have considered until now the so-called *Cartesian complexes* which are Cartesian products [4] of one-dimensional complexes. A one-dimensional complex is a sequence of interchanging incident points and cracks.

It is possible to introduce in Cartesian complexes coordinates of cells. A one-dimensional complex being the *X*-axis of the complex is a sequence of interchanging 0- and 1-cells. They can be numerated by subsequent integer numbers as shown in Fig. 2.7.

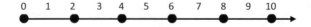

Fig. 2.7 A one-dimensional complex with coordinates of its cells

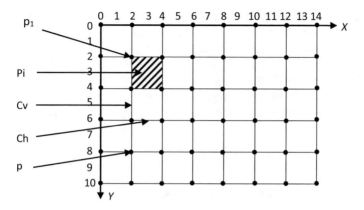

Fig. 2.8 Combinatorial coordinates in a two-dimensional ACC

In a two-dimensional Cartesian complex either its X-axis or its Y-axis are one-dimensional complexes. Each cell in a column of the two-dimensional complex obtains its x-coordinate equal to the number of the cell of the X-axis in this column. Similarly, each cell in a row of the complex obtains its y-coordinate equal to the number of the corresponding cell in the one-dimensional complex of the Y-axis. We call these coordinates *combinatorial coordinates of cells*. We consider here a coordinate system with the Y-axis directed downwards. It is easily possible to use our considerations also for the system with the Y-axis directed upwards as it will be explained below.

Consider please the Fig. 2.8 above. The pixel "Pi" has the coordinates (3, 3), and it is easily seen that all pixels have both coordinates odd.

The vertical crack "Cv" has the coordinates (2, 5), and each vertical crack has an even x and an odd y coordinate.

The horizontal crack "Ch" has the coordinates (3, 6), and each horizontal crack has an odd x and an even y coordinate.

The point "p" has coordinates (2, 8), and each point has both coordinates even.

We consider mostly a cell complex corresponding to an original digital image of the size width* × height. The grid of the complex must have 2*width + 1 columns and 2*height + 1 rows. The additional column and the additional row are necessary because we mostly must consider cells at the right-hand side and the bottom of the complex. We call the coordinates of cells in a complex *combinatorial coordinate*, while the pixels in the original digital image have other coordinates. The original grid of Fig. 2.8 has 7 columns and 5 rows with pixels. Thus, the original image was a 7 × 5 image, and the pixel "Pi" has in the original image the coordinates (1, 1). We call the original coordinates *standard coordinates*. Their relation to the combinatorial coordinates is amazingly simple: the standard coordinate is the result of the integer division of the combinatorial coordinate by two.

We shall use in our algorithms both combinatorial and standard coordinates. Points and cracks have combinatorial coordinates different from the coordinates of an adjacent pixel. Thus, the point p_1 in the upper left corner of the pixel Pi (Fig. 2.8

above) having combinatorial coordinates $(3, 3)$, has combinatorial coordinates $(2, 2)$. However, the standard coordinates of both Pi and p_1 are the same, namely $(1, 1)$. Similarly, the crack C_1 on the left side of the pixel Pi has combinatorial coordinates $(2, 3)$. However, the standard coordinates of both Pi and C_1 are the same, namely $(1, 1)$. Thus, it is impossible to distinguish between these three cells by their standard coordinates. They should be distinguished by their dimensions. Nevertheless, there are cases in applications in which we can use standard coordinates for points or cracks as we will see below.

2.3   Boundaries in Cell Complexes

Since the abstract cell complexes satisfy all axioms of the classical topology, an abstract cell complex is a special case of a classical topological space. It is the so called Alexandroff space [5] after the Russian scientist P. Alexandroff who together with H. Hopf published 1935 a work [6] about the topology of complexes. It is simultaneously a Kholmogorov space [7]. An abstract cell complex satisfies the separation axiom T_0. Thus, we can use the classical definition of a frontier:

The frontier of a subset S of an ACC is the set of all cells whose smallest neighborhood contains both cells belonging to S and cells not belonging to S.

The notion of "frontier" is widely used in topological publications. The difference between the notions of frontier and boundary is the following: Frontier of a subset S of the space R is the set of all cells whose smallest neighborhood crosses the subset S and its complement $R - S$. Boundary of the two-dimensional subset S is the closure of the set of all 1-cells of R whose smallest neighborhood contains exactly one 2-cell of S. This difference is less important for our presentation. Therefore, we write everywhere "boundary" for "boundary or frontier".

Thus, a boundary can contain no pixels since the smallest neighborhood of a pixel contains a single cell: the pixel itself, and cannot contain a cell of S *and* another cell not belonging to S. The boundary in a cell complex is thin, as in the classical topological space. Its area is equal to zero.

The boundary is a subset of a two-dimensional ACC containing only points and cracks. A crack C belongs to the boundary of the set S if among the three cells in the smallest neighborhood of C there are cells belonging to S and cells not belonging to S. For example, one pixel of the smallest neighborhood of C is in S and the other is not in S. Or another example: The crack C lies at the border of the complex. C does not belong to S, and C is incident to a pixel of S.

A point P belongs to the boundary of S if the smallest neighborhood of P is "mixed", i.e., it contains cells both belonging to S and not belonging to S. Figure 2.9 shows an example of a boundary.

Figure 2.9a shows a complex. Only the hatched squares, the bold-drawn cracks and the marked point belong to the subset S. Figure 2.9b shows the boundary of S.

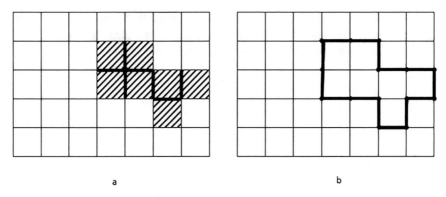

Fig. 2.9 Example of a complex (**a**) and its boundary (**b**)

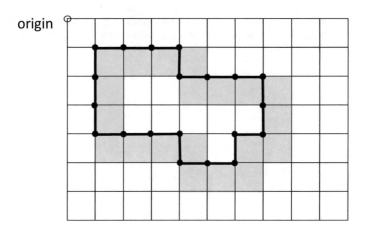

Fig. 2.10 Boundary presented by shaded pixels assigned to points and cracks of the boundary

2.4 Graphical Presentation of Boundaries

A boundary in a cell complex consists of points and cracks. The hardware of a computer delivers no possibility for presenting these cells graphically. If it is possible to magnify the image, then it is possible to present the cracks as vertical and horizontal lines with the method "DrawLine".

However, there is also another universal possibility usable without magnification. It is possible to uniquely assign a pixel to each point and to each crack. For example, of the two pixels in the smallest neighborhood of a crack the pixel lying farther from the origin (0, 0) of the coordinate system is assigned to the crack. Similarly, of the four pixels about a point the pixel which is farther from the origin than each of the other three pixels, is assigned to the point. The assigned pixels can be presented graphically. Figure 2.10 shows the presentation of the boundary as the set of shaded pixels assigned to the points and cracks of the boundary.

It is rather astonishing that H.Y. Feng and T. Pavlidis [8] have suggested this presentation of the boundary, which they called "extended boundary", about 40 years ago without saying a word about cell complexes.

References

1. Listing J. Der Census räumlicher Complexe. Abhandlungen der Königlichen Gesellschaft der Wissenschaften zu Göttingen. 1862;10:97–182.
2. Hazewinkel M, editor. Hausdorff space. In: Encyclopedia of mathematics. Springer/Kluwer Academic Publishers; 1994/2001. ISBN 978-1-55608-010-4.
3. Whitehead JHC. Combinatorial homotopy. I. Bull Am Math Soc. 1949;55(5):213–45.
4. Warner S. Modern algebra. Dover Publications; 1990. p. 6.
5. Alexandroff P. Discrete spaces. Mat Sbornik. 1937;2:501–18.
6. Alexandroff P, Hopf H. Chapter 3: Topologie der Komplexe. In: Topologie II. Chelsea; 1935.
7. Aleksandrov PS. Chapter XVIII: topology. In: Aleksandrov AD, Kolmogorov AN, Lavrent'ev MA, editors. Mathematics/its content, methods and meaning. 2nd ed. MIT Press; 1956/1969.
8. Feng HY, Pavlidis T. IEEE Trans Circ Syst. 1975;22:427–39.

Chapter 3
Boundary Tracing in Binary Images Using Cell Complexes

Abstract This chapter gives a detailed description of the tracking of boundaries in a binary image because the case of a binary image is the simplest one, and it can serve as a basis for the cases of indexed and color images. The chapter describes the rules for defining the direction of the next step depending on the colors (black or white) of the two pixels in front of the movement along the boundary. The Sect.3.1 contains the full subscription of algorithms "SearchBin" and "TraceBin".

Keywords Boundary tracing · Binary image · Directions of cracks · Turning rule · Algorithms · Components

To realize the tracing of the boundary of a subset in a digital image it is advisable and recommendable to define an abstract cell complex related to the image.

Consider first the simplest case of tracing *boundaries in a binary image* in which there are no black pixels at the borders of the image. Let the set S be the set of black pixels. We consider here the coordinate system with the positive Y-axis directed downwards. We call it the graphical coordinate system to distinguish from the mathematical coordinate system where the positive Y-axis is directed upwards.

The tracing will be realized by the algorithm "SearchBin" which scans the image line by line while starting with a pixel in the upper left corner of the image and looking for a black pixel whose left neighbor pixel is white. When such a pair of pixels is found, then the algorithm checks whether the black pixel is labeled as already visited. If it is not labeled, then the algorithm starts the *tracing algorithm* with the coordinates of the black pixel as parameters. The tracing algorithm must define variables corresponding to cells of a cell complex: points, cracks, and pixels. These variables must have combinatorial coordinates (see above).

The tracing subroutine can start either downwards (direction $= 1$) or upwards (direction $= 3$). We decide in this simple example for upwards. In this case the tracing of the boundary will be performed in such a way that *black pixels are always at the right side* of the tracing. Later, we will examine all possible variants of the tracing algorithm.

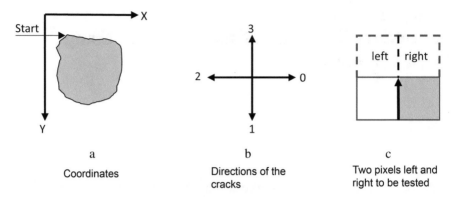

Fig. 3.1 Finding the starting crack (a), directions of the cracks (b) and two pixels "left" and "right" that must be tested (c)

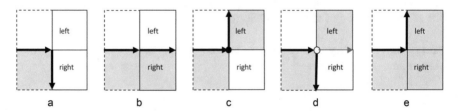

Fig. 3.2 Decisions about the direction of the next step after direction 0; explanation in text

Tracing the boundary of a connected subset of pixels having all the same color must be performed in such a way that this color remains always on the same side of the directed steps along the boundary. To make the next step, the subroutine must check the two pixels "left" and "right" in front of the directed crack (Figure 3.1c above). The tracing algorithm takes a decision about the direction of the next step according to the colors of these two pixels as explained in Fig. 3.2 above. We consider first the tracing of the boundary of a black region while black pixels must be always on the right side of directed steps. Figure 3.2 shows the simplest case of the situation after a step along the boundary has been made in the horizontal direction to the right (direction = 0).

In Fig. 3.2 the step already made in direction 0 is shown as the horizontal arrow at the left side. Figure 3.2a shows the case when both pixels "left" and "right" have white color. In this case the crack directed downwards is a boundary crack and is the only one of the three cracks that could serve as a successor to the step already made. In the case of Fig. 3.2a the movement of the crack must turn to the right.

In the case of Fig. 3.2b the pixel "left" is white while the pixel "right" is black. The crack directed to the right is the only boundary crack and the direction does not change; the next step goes in direction 0.

In the case of Fig. 3.2c the pixel "left" is black while the pixel "right" is white. This is a special situation: the zero-dimensional cell in the middle is singular. This

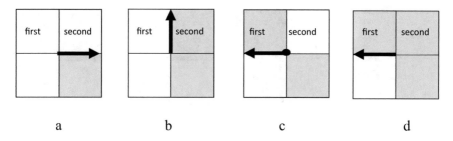

a   b   c   d

Fig. 3.3 Directions of the first step depending on the colors of the two upper pixels

means that the set of black cells bounded by this cell is disconnected. All three cracks, which can serve as successors of the last step, belong to a boundary. The horizontal one of these three cracks is not acceptable because the traced black color lies on its wrong side.

The decision about the choice of one of the two remaining cracks depends upon the decision of whether the singular zero-dimensional cell belongs to the black or to the white subset. This is the same as which of the two diagonally adjacent pairs of the pixels, the white one or the black, is connected. The movement along the boundary should not cross a connected subset. According to the suggestion of Rosenfeld [1], the black subset in a binary image is connected. We accept this suggestion for our simplest case. In Chap. 4 we will explain how to decide about the membership of a singular zero-dimensional cell.

Similar decisions should be made in situations after a step along the boundary has been made in other directions. All these decisions in the case of tracing the boundary of a black region, when black pixels are on the right side, can be made according to the following *turning rule* which is guilty for binary, indexed and color images.

If the pixel "left" and the central zero-cell belong to the black, then turn to the left. Otherwise, if the pixel "right" and the central zero-cell belong to the white, then turn to right. In the remaining case retain the direction.

Now it is necessary to decide which should be the direction of the first step of the algorithm "TraceBin". The decision depends on the colors of two upper pixels "first" and "second" about the starting point as shown in Fig. 3.3 The lower two pixels are the pixels found by the algorithm "SearchBin" as a *black* pixel following a white one. The starting point lies in the middle of the four pixels.

Let us consider all four combinations of the colors of two upper pixels (Fig. 3.3).

We consider first the case in which both upper pixels belong to the white (Fig. 3.3a). We look for such a direction of the first step of the tracing that the white pixels are on the left and the black ones on the right side. In this case the direction is 0 (horizontal to the right).

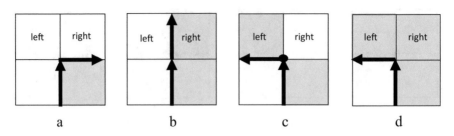

a   b   c   d

Fig. 3.4 Decisions about the direction of the next step after direction 3

If the first of the upper pixels is white and the second one is black (Fig. 3.3b), then the step upwards (direction 3) lies between a white pixel on the left and a black pixel on the right side. The direction is 3.

If the first of the upper pixels is black, then, independently of the color of the second pixel (Fig. 3.3c, d), the step to the left (direction 2) is correct because, as explained above, the starting point belongs to the black subset. The direction is 2.

It is easily seen that these directions correspond to the turned directions according to the above turning rule for the direction 3 (Fig. 3.4 above) because direction 3 is the only direction for which the pixels "left" and "right" coincide with the upper two pixels about the starting point.

Therefore, it is rational, instead of formulating the rules for the choice of the direction of the first step, to define the starting direction equal to 3 and use the turning rule at the start of the algorithm "TraceBin".

However, in all four cases shown in the above Fig. 3.3 a step downwards (direction 1) would be a step along the boundary. But in this case the black pixel would be on the wrong side. To solve this problem, it is necessary to choose a preferred color depending on the starting direction and make the tracing in such a way that the preferred color, instead of black color, lies on the right side. Thus, the starting direction can be chosen either as 3 or as 1, the preferred color being black for starting direction 3 and being white for starting direction 1. The above formulated turning rule must be reformulated as follows:

If the pixel "left" has the preferred color, then turn to the left. Otherwise, if the pixel "right" has not the preferred color, then turn to right. In the remaining case retain the direction.

If the user wants to save the traced boundary, then it is necessary to write the values of the subsequent directions into an array. Let us call it "Directions". It is also necessary to write down the address of the last direction of each component of the boundary saved in "Directions". It is not necessary to write down the address of the first direction of the component since it can be calculated as the address of the last direction of the previous component plus 1.

The searching algorithm should write down the coordinates of the starting point of each component. Thus, three values should be saved for each component of the boundary: two coordinates of the starting point and the address of the last direction. The latter address is the index of the last directory in the array "Directions". These three values can be put together in the structure "Loop" containing two coordinates of the starting point and one integer value of the address of the last directory of each component of the boundary. We shall call a component of the boundary a loop since in a boundary of a binary image all components are closed curves.

Thus, the user will obtain after the processing of the whole image two arrays: an array of "Loops" and another array "Directions" of the directions of the cracks of all components of the boundary.

Now we describe two simplest algorithms "SearchBin" and "TraceBin" for tracing boundaries in a binary image containing no black pixels at the borders of the image. We shall consider the more general version without this limitation later. The tracing must retain black pixels at its right-hand side. The singular points should be considered as belonging to the subset of black pixels.

We are using the graphical coordinate system (positive Y-axis directed downwards). Remember that it is necessary to label visited vertical cracks. This can be made by the algorithm "TraceBin" if there is an array containing a memory location for each of the vertical cracks. This array must have such a size that there would be place for each cell of the cell complex. Thus the array for the cell complex must have the size of $(2*width + 1)*(2*height + 1)$ elements. The addition "+ 1" is necessary because points and cracks that lie at the extreme right and the extreme bottom border of the image should be considered. Both algorithms should use variables representing cells of all three dimensions: points, cracks, and pixels.

There is a simpler solution: it is possible to label instead a vertical crack the pixel having the same standard coordinates as the crack. Then the array "Label" for labeled cells would have the size (width*height) and only the algorithm "TraceBin" should have variables representing cells of all three dimensions. However, in this case it is important to exactly define the coordinates of the pixel which should be labeled. This pixel has the combinatorial coordinates (Crack.X + 1, Crack.Y) and its standard coordinates equal ((Crack.X + 1)/2, Crack.Y/2). In the case when the starting direction of "TraceBin" equals 3 (upwards) it would be not correct to set the pixel for labeling being equal to the point. We decide to describe the slightly more complex version using the array "Label" of the size $(2*width + 1)*(2*height + 1)$ and using variables for cells of all three dimensions.

The algorithms "SearchBin" and "TraceBin" have access to the two-dimensional arrays "Image[width, height]" and "Label[(2*width + 1), (2*height + 1)]". Please note that the binary "Image" should be an array of simple pixels each containing a single value "black" or "white" and not three channels "red", "green" and "blue" as in a color image.

The algorithm "SearchBin" scans the vertical cracks of the cell complex representing the image. The cell complex has the sizes represented with the variables

"C_width" and "C_height" but it is explicitly represented only by the array "Label" which is used for labeling already visited vertical cracks. This is necessary to prevent that a component of the boundary be traced many times. The variable "Crack" being a two-dimensional vector with the components X and Y which are combinatorial coordinates of the crack, is explicitly present in the algorithm. The standard coordinates of the two-dimensional vector "Pixel", which is also explicitly present, are calculated from coordinates of "Crack" and used for directly getting the values of the image. The image is presented, as usual, in standard coordinates.

The algorithm "SearchBin" scans the vertical cracks row by row, calculates standard coordinates of "Pixel": Pixel.X=(Crack.X+1)/2, Pixel.Y=Crack.Y/2 and looks for a black pixel "Image[Pixel.X, Pixel.Y]" following a white pixel "Image [Pixel.X - 1, Pixel.Y]". It checks whether "Label[Crack.X, Crack.Y]" is equal to 1 which is an indication, that the corresponding crack has been already visited. If this is not the case, the algorithm "TraceBin" is started with the coordinates of the crack as parameters.

To save the results of tracing the boundaries we use the array "Loop" of structures and the array "Directions" where the algorithm "TraceBin" called by "SearchBin" saves the directions of subsequent cracks of a loop. "Directions" is an array of small numbers. Each "Loop" contains the standard coordinates of the two-dimensional vector "Start" representing the starting point of the "Loop" and the index of the last direction of the loop. The first direction of a loop must not be saved since it can be easily calculated as the last index of the previous loop plus 1. The first index of the fist loop is obviously equal to 0. Thus, the array "Loop" is an array of structures of the type "SLoop":

```
structure SLoop
{
2D vector Start;
integer Last;
}.
```

3.1   Algorithm "EncodeBin"

We describe first as introduction the algorithm "EncodeBin" for tracing and encoding boundaries in binary images. It consists of two parts: "SearchBin" and "TraceBin".

To make the algorithms slightly simpler it is rational to declare the number "nLoop" of detected components (loops) of the boundary, the number "nDir" of the saved directions, the array "Loops[maxLoop]" and its size "maxLoop" in such a way that all they are accessible to both algorithms "SearchBin" and "TraceBin". Also, the array "Directions[]" and its size "maxDir" should be declared in this way. We call such variables and arrays "global".

It is important to note for all algorithms that variables not declared in the algorithm are global and can be used in other algorithms.

Algorithm "SearchBin(Image)" for binary images:

Step 1: Declare and initialize global variables maxLoop ← 10000,
   maxDir ← 100000, nLoop ← 0, nDir ← 0.
Step 2: Declare global arrays Loop[maxLoop] and Directions[maxDir].
Step 3: Declare variables C_width, C_height.
Step 4: Initialize variables
   C_width ← 2*(width of the Image) + 1.
   C_height ← 2*(height of the Image) + 1.
   nLoop ← 0.
Step 5: Declare the two-dimensional array Label[C_width, C_height].
Step 6: Set all elements of Label equal to 0.
Step 7: Declare variables Crack and Pixel as two-dimensional vectors.
Step 8: Initialize variable Crack.Y ← 1.
Step 9: Repeat the steps until Crack.Y is less than C_height:
  9.1: Set Pixel.Y ← Crack.Y /2.
  9.1: Initialize variable Crack.X ← 2.
  9.2: Repeat the steps until Crack.X is less than C_width:
  9.2.1: Set Pixel.X ← (Crack.X + 1)/2.
  9.2.2: If Image[Pixel.X - 1, Pixel.Y] equals white and
    Image[Pixel.X, Pixel.Y] equals black
    and Label[Crack.X, Crack.Y] equals 0. then
    Begin
     Set Loop[nLoop].Start.X ← Crack.X /2.
     Set Loop[nLoop].Start.Y ← (Crack.Y - 1)/2.
     Start TraceBin(Image, Crack, Label).
     Set Loop[nLoop].Last ← nDir. (nDir is changed by TraceBin)
     Set nLoop ← nLoop + 1.
    End   (End if Image[Pixel.X, ...])
  9.2.3: Set Crack.X ← Crack.X + 2. (End step 9.2)
  9.3: Set Crack.Y ← Crack.Y + 2. (End step 9)
Stop.

The algorithm "TraceBin" obtains as parameters the image, the variable "Crack" representing the starting crack and the array Label. The variable *nDir* which is the number of already saved directions and the array "Directions" are accessible to "TraceBin". The value *nDir* must be submitted in such a way that the algorithm "TraceBin" can change it.

The algorithm "TraceBin" calculates combinatorial coordinates of the starting point as that of the upper point of the starting crack which has a predefined direction 3 (upwards). The moving "Point" is set equal to the starting point. Then "TraceBin" defines the color of the pixels which must lie at the left side of the traced boundary. Let us call this color the *traced color*. In our special case this is the color at the left side of the starting crack. The traced color could be defined by the algorithm

"SearchBin" described above as "white". However, we prefer to calculate this color in "TraceBin" for the case that "TraceBin" will be used with another version of "SearchBin".

The algorithm "TraceBin" starts then the repetition of a set of instructions containing instructions for moving "Crack" and "Point" until "Point" reaches again the starting point. Thus "Point" runs along a closed loop. "TraceBin" calculates in the loop coordinates of the pixels "Left" and "Right" lying before the directed crack, specifies values of "Image" at these two pixels after converting them to standard coordinates. Remember please that "Image" is given in standard coordinates. The direction of the next step of the tracing is specified by comparing the values of "Left" and "Right" with the traced value.

We are using in the algorithm "TraceBin" variables representing two-dimensional vectors. For example, the variable "Point" is a two-dimensional vector with components "Point.X" and "Point.Y". It represents a zero-dimensional cell.

The algorithm "TraceBin(Image, Crack, Label)" needs three small arrays "toLeft", "toRight" and "Step" of two-dimensional vectors with constant components. Each of these arrays contains four vectors, one vector for each of the four possible directions: 0, 1, 2 and 3. The arrays "toLeft" and "toRight" are necessary to calculate coordinates of two pixels in which the value of the image should be estimated with the aim to decide about the direction of the next step of the tracing (compare Fig. 3.4 above). The array "Step" is used to calculate the coordinates of a point or a crack making a step in the new direction. These arrays have the following elements:

```
toLeft[0] = (1, -1); toLeft[1] = (1, 1); toLeft[2] = (-1, 1); toLeft[3] =
(-1, -1);
toRight[0] = (1, 1); toRight[1] = (-1, 1); toRight[2] = (-1, -1); toRight
[3] = (1, -1);
Step[0] = (1, 0); Step[1] = (0, 1); Step[2] = (-1, 0); Step[3] = (0, -1);
```

Algorithm TraceBin(Image, Crack, Label) for binary images:

The fields "toLeft", "toRight" and "Step" are described in the above text.

Step 1: Declare variables value, startDir and dir.
Step 2: Declare two-dimensional vectors Point, Crack, Pixel, startPoint, Left, and Right.
Step 3: Initialize variables:
    startPoint.X ← Crack.X,
    startPoint.Y ← Crack.Y – 1,
    Point ← startPoint,
    startDir ← 3, and
    dir ← startDir.
Step 4: If startDir equals 1, then set value ← Image[(x - 1)/2, y/2].
        else set value ← Image[x/2, y/2].
Step 5: Repeat the steps if Point is not equal to startPoint:
    5.1: Set Left ← Point + toLeft[dir].
    5.2: Set Right ← Point + toRight[dir].
    5.3: If Image[Left.X/2, Left.Y/2] is equal to value, then set dir ← dir+3.
         else
            if Image[Right.X/2, Right.Y/2] is not equal to value, then set dir ← dir+1.
    5.4: If dir is greater than 3, then set dir ← dir - 4.
    5.5: Set Crack ← Point + Step[dir].
    5.6: If dir is equal startDir, then set Label[Crack.X, Crack.Y] ← 1.
    5.7: Set Directions[nDir] ← dir.
    5.8: Set nDir ← nDir + 1.
    5.9: Set Point ← Point + Step[dir].   (End repeat step 5)
Stop.

Let us consider an example. Figure 3.5 shows a small binary image and its detected boundary components.

The data of the found components (loops) of the boundary are shown in Table 3.1.

The data structure describing all boundaries of the image contains two arrays: the array of loops and the array of directions. However, this date structure is not

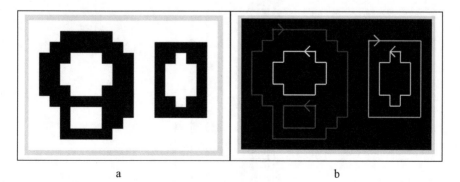

a   b

Fig. 3.5 Example of a binary image (**a**) and its detected boundary components (**b**)

Table 3.1 Data of the five components of the boundary of the image of Fig. 3.5a

Loop	Start	First	Last	Directions
0	(3, 1)	0	37	0;0;0;0;1;0;1;0;1;1;1;1;1;1;1;2;1;2;1;2;1;2;2;2;2;3;3;3;2;3;2;3;3;3;0;3;0;3.
1	(12, 2)	38	61	0;0;0;0;1;1;1;1;1;1;1;2;2;2;2;3;3;3;3;3;3;3.
2	(7, 3)	62	79	2;2;2;1;2;1;1;0;1;0;0;3;0;3;3;2;3.
3	(15, 3)	80	95	2;1;2;1;1;0;1;0;3;0;3;3;3;2;3.
4	(7, 8)	96	105	2;2;2;1;1;0;0;0;3;3.

complete: it is for example not possible to find from this structure the boundary of the left black subset of Fig. 3.5a which consists of loops 0 (color red in Fig. 3.5b), loop 2 (color yellow) and loop 4 (color dark green) because it does not indicate that the loops 2 and 4 lie inside the loop 0. Also, it is not possible to find from this structure the boundary of the great white subset including both black subsets. This boundary consists of the great loop containing the whole image and the loops 0 and 1 (colors red and orange in Fig. 3.5b). The great loop containing the whole image is even not present in the produced structure which we have presented because it is so simple and easy to understand.

To find a complete boundary of a subset containing holes, like the left black subset of Fig. 3.5a, it is necessary to develop an algorithm which would find that the boundaries of holes of a subset lie inside the outer boundary of the subset. We shall describe this algorithm "MakeTree" below.

Reference

1. Rosenfeld A. Connectivity in digital pictures. J ACM. 1970;17:146–60.

Chapter 4
Boundary Tracing and Encoding in Color Images

Abstract This chapter describes two main algorithms for encoding boundaries in color or indexed images. The main algorithm CORB is simple and that's why people like to use it. The main algorithm "MakeLineList" is more complicated and brings as the result a more economical code. The chapter contains also a simple algorithm "TraceEqaution" for tracing a curve defined by means of an equation.

Keywords Color and indexed images · EquNaLi rule · CORB algorithm · Reconstruction · Positive and negative loops · Containment tree · Tracing equations · MakeLineList algorithm · Restoration of encoded images

We have described in the previous chapter the tracing and encoding of boundaries in binary or black and white images. Here we describe algorithms usable in the general case of color or indexed images with many colors. The algorithm CORB traces the boundary of a connected subset whose all pixels have a fixed value. We shall call these values *the colors* even in the cases when the values are some abstract numbers, e.g., the indices of colors of an indexed image. This fixed color is regarded as foreground while all other colors are regarded as background. Thus, the algorithm CORB considers the image during the tracing of one boundary as a binary image. A boundary in a binary image is always a closed line. This makes the algorithm CORB so simple.

However, this algorithm has a drawback: Each boundary crack is visited twice: for the first time as a crack of the boundary of a hole in a region with the fixed color and for the second time as a crack of the boundary of a region contained in the hole. To fix this drawback we suggest in the Sect. 4.3 the algorithm "MakeLineList" which visits each boundary crack only ones. At the price of this benefit, the algorithm is much more complicated.

The rules of the choice of the direction of the next step during the tracing are like that for binary images (Chap. 3) with an essential difference: In the case when the zero-dimensional cell at the start of the next step is singular, it is possible to choose the direction of the next step by means of a method investigating the membership of pixels about the singular cell.

4.1   Rules of the Choice of the Direction of the Next Step

The tracing algorithm takes a decision about the direction of the next step according to the colors of two pixels "left" and "right" in front of the actual step as explained in Fig. 4.1 below. We consider first the tracing of the boundary of a white region while white pixels must be always on the left side of directed steps. White is now the fixed color: the tracing lets the fixed color on the left side. Figure 4.1 shows the simplest case of the situation after a step along the boundary has been made in the horizontal direction to the right (direction = 0).

In Fig. 4.1 the step already made in direction 0 is shown as the horizontal arrow at the left side. Figure 4.1a shows the case when both pixels "left" and "right" have the fixed color (white). In this case the crack directed downwards is a boundary crack and is the only one of the three cracks that could serve as a successor to the step already made. In the case of Fig. 4.1a the movement of the crack must turn to the right.

In the case of Fig. 4.1b the pixel "left" has the fixed color while the pixel "right" has some other color. The crack directed to the right is the only boundary crack and the direction does not change; the next step goes in direction 0.

In the case of Fig. 4.1c the pixel "right" has the fixed color while the pixel left has some other color. This is a special situation: the zero-dimensional cell in the middle is *singular*. This means that the set of cells with the fixed color bounded by this cell is disconnected. All three cracks, which can serve as successors of the last step, belong to a boundary. The horizontal one of these three cracks is not acceptable because the fixed color to be traced lies on its wrong side. The decision about the next step in the case of a singular central point can be made by means of the special turning rule called "EquNaLi rule" defined below. If the EquNaLi rule reports that the singular cell belongs to the set with the other as fixed color, then the diagonal pair with this color becomes connected and the movement turns to the left (Fig. 4.1c).

If, however, the EquNaLi rule reports that the singular cell belongs to the set with the fixed color, then the diagonal pair with this color becomes connected and the movement turns to the right (Fig. 4.1d).

In the case of Fig. 4.1e both pixels "right" and "left" have other than fixed color. In this case the movement turns to the left side.

The decision about the choice of one of the two remaining cracks depends upon the decision of whether the singular zero-dimensional cell belongs to the subset with

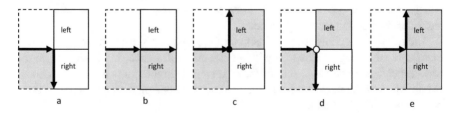

Fig. 4.1 Decisions about the direction of the next step after direction 0; explanation in text

the fixed color. Now we will explain how to decide about the membership of a singular zero-dimensional cell.

The membership of a singular zero-dimensional cell can be correctly defined by means of the "EquNaLi" rule described below. This rule has been suggested by the author many years ago [1] and was successfully used in projects developed by the author.

Two pixels with coordinates (x, y) and $(x + 1, y + 1)$ or (x, y) and $(x - 1, y + 1)$ compose a diagonal pair. A diagonal pair of adjacent pixels is connected if both pixels have the same color and the zero-dimensional cell lying between them has also this color. It has, naturally, little sense to speak of the color of zero-dimensional cells because their area is zero. However, it is necessary to use this notion to consistently define connectedness in digital images. To avoid using the notion of colors of zero-dimensional cells we suggest replacing "color" by "label" while the label of a pixel is equal to its color.

EquNaLi Rule for two-dimensional images:
The label of a singular zero-dimensional cell c^0 is defined as follows:

If the 2×2 neighborhood of c^0 contains exactly one diagonal pair of pixels with equal labels ("Equ"), then c^0 receives this label. If there is more than one such pair but only one of them belongs to a narrow ("Na") stripe, then c^0 receives the label of the narrow stripe. Otherwise c^0 receives the lighter ("Li") label (i.e., the greater label) of the pixels in the neighborhood of c^0.

The latter case corresponds to the cases when the neighborhood of c^0 contains two diagonal pairs with equal labels and both belong to a narrow stripe.

To decide whether a diagonal pair about c^0 belongs to a narrow stripe it is necessary to scan an array of $4 \times 4 = 16$ pixels with c^0 in the middle and to count the labels corresponding to both diagonals. The smallest count indicates the narrow stripe.

Consider the example of the following Fig. 4.2. Suppose that it is necessary to define the connectedness in such a way that both the white and the black "V" are connected. This may be important, since a disconnected subset would be disintegrated to many small components. This is obviously impossible under any adjacency relation. The aim can be achieved by means of the above EquNaLi rule assigning membership labels (e.g., integers corresponding to colors) to zero-dimensional cells.

4.2   Encoding Boundaries with the CORB Algorithm

An image is a set of pixels with some values assigned to its pixels. A homogeneous region of the image is a connected subset of pixels whose all pixels have the same value. Each region has its outer boundary. A region *Reg* can have one or more inner boundaries each corresponding to a hole in *Reg*. Thus, for example, the region R_3 in

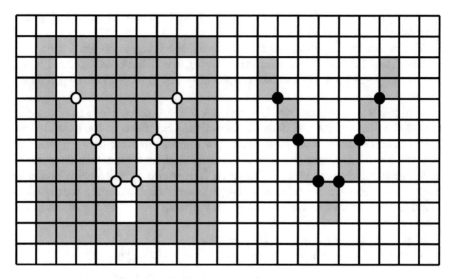

Fig. 4.2 A white and a black V-shaped regions in one image; both are connected due to applying the EquNaLi rule

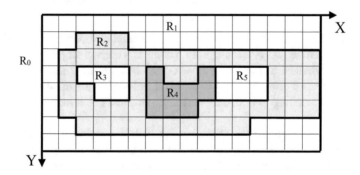

Fig. 4.3 Example of a multivalued indexed image (coordinate system of computer graphics)

Fig. 4.3 is a hole in the region R_2. A hole in turn can contain one or more other regions, e.g., the regions R_4 and R_5 are contained in a hole of R_2 (Fig. 4.3).

We describe below an improved main algorithm which we call the "Components of Region Boundaries", abbreviated as CORB. It consists of four sub-algorithms: "Search", "Trace", "MakeTree" and "Restore".

The aim of the sub-algorithm "Trace" is to trace and encode each component of the boundary of each region of a 2D image. An especial property of the algorithm CORB is that the tracing of a boundary component chooses one of two colors on the sides of the boundary as the "foreground color" and considers all other colors as the background. The tracing is performed in such a way that the foreground color remains always at the same side of the tracing. Thus, the image is from the view of tracing a *binary image* with two colors: foreground and background, while background is any color different from the foreground color. Correspondingly the boundary becomes a closed curve, as all boundaries in a binary image. This is an important advantage of the CORB algorithm, and it makes the algorithm so simple.

Consider an example: the region R_2 of the image of Fig. 4.3 contains a hole, which is another region R_3 of the image. Then the boundary of R_3 will be traced and encoded twice: first as a hole in R_2 and as a component of the boundary of R_2 with the color of R_2 as foreground color and then as the outer boundary of the region R_3 with the color of R_3 as foreground color. In a binary image being regarded as a cell complex each component of the boundary is a closed sequence of points and cracks.

The algorithm finds a boundary crack, i.e., a crack incident to two pixels with different colors. The algorithm checks whether this crack was labeled as already visited. If it was not visited, then the algorithm fixes one of these colors as the foreground color and starts the tracing procedure which must trace a boundary component of a subset having the foreground color. The starting point of the tracing is one of the end points of the found crack. The algorithm saves for each curve (which we call a "loop") the coordinates of its starting point (either standard or combinatorial coordinates, no matter) and the directions of all oriented cracks of the boundary. The cracks which are visited during the tracing must be labeled as "already visited" so that no one of these labeled cracks can be used to start the tracing procedure again. The coordinates of the starting point and the directions of all cracks of the boundary curve saved for all boundary curves of the image are sufficient for an exact reconstruction of the original image in the way which will be explained below.

To reconstruct the original image from the codes of the boundaries it is also necessary to encode the outer boundary of the whole image. For this purpose, we consider a virtual outer area R_0 (Fig. 4.3 above) having a non-existing color, e.g., -1. Then the regions R_1 and R_2 of Fig. 4.3 are regarded as components inside of R_0. The components of the boundaries of the regions of the whole image compose a structure of an *inclusion tree* which shows which components are contained in the interior of some other component.

We describe below the algorithms CORB for tracing and encoding boundaries in color or indexed images and for restoring the original image from the arrays Loop and Directions. The variable "color" can represent either the color or the index contained in a pixel of the image. The algorithm CORB consists of four sub-algorithms: "Search", "Trace", "MakeTree" and "Restore".

The sub-algorithm "Search" scans the image row by row and looks for a pair of subsequent pixels with different colors. If the crack lying between these two pixels was not already visited, then the algorithm fixes the color of the first of these two pixels and calls the algorithm "Trace" with the fixed color as the foreground color and the coordinates of the starting point as arguments.

In the following algorithm "Search" the notation "Image[width, height]" stands for a two-dimensional array whose elements contain colors or indices in the case of an indexed image. The values "width" and "height" are the numbers of columns and rows in the array "Image[,]" respectively. Similarly, "Lab[width + 1, height]" is a global two-dimensional array of the size $((width + 1) \times height)$. The width of "Lab" is increased by 1 since it is necessary to label cracks at the right-hand side of the image. The constant "Out" is the virtual, non-existing color of pixels outside the image.

We want making the algorithms applicable for both color and indexed images. Therefore, we use the subroutine "ToInteger", which transforms a color consisting of three channels Red, Green, and Blue to an integer value: The value of the Red channel is shifted by 16 bits to the higher values; the value of Green by 8 bits; the value of Blue remains not shifted. All three shifted values are joined in a single integer value. If "Image[x0, y0]" is an index, then "ToInteger" lets it unchanged. The variable "nLoop" of the following algorithm "Search" defines the number of already detected loops and the variable "nDir" defines the number of already detected directions of the cracks in the loops. These variables must be saved in such a way that they can be accessible to the algorithm "Restore" described below. The values of these variables are submitted to the algorithm "Trace" as parameters.

Algorithm Search(Image[]):

Step 1: Declare the variables nDir, nLoop, NX, NY, value, oldValue, Out,
      x0 and y0.
Step 2: Set NX ← Image.width and NY ← Image.height and Out ← -1.
Step 3: Set elements of Lab equal to 0.   (Array of labels "already visited")
Step 4: Initialize nLoop ← 0 and nDir ← 0.   (Indices of loop and direction.)
Step 5: Set y0 ← 0.
Step 6: Repeat the steps until y0 is less than NY:
    6.1: Set x0 ← 0 and oldValue ← Out.
    6.2: Repeat the steps until x0 is less than NX:
    6.2.1: If x0 is less than NX, then set value ← ToInteger(Image[x0, y0]).
       else set value ← Out
    6.2.2: If value is not equal to oldValue and Lab[x0, y0] is equal to 0, then
                 call Trace(Image, x0, y0, nLoop, nDir, oldValue).
    6.2.3: Set nLoop ← nLoop + 1.
    6.2.4: Set oldValue ← value.
    6.2.5: Set x0 ← x0 + 1.   (End repeat 6.2.)
    6.3: Set y0 ← y0 + 1.   (End repeat 6.)
    }
Stop.

The algorithm "Trace" runs along the boundary and makes records into the structure "Loop[iLoop]" of the actual loop and into the array "Directions", where all subsequent directions of the boundary cracks are stored. The structure of a loop (slightly different from the structure "SLoop" mentioned in Chap. 3) contains the starting point as a two-dimensional vector, the integer foreground color assigned to the loop and the index "Last" in the array "Directions" of the direction of the last crack of the loop. It is not necessary to store the index of the first direction of the loop since it can be easily computed: The first index for Loop[0] is equal to 0 and the first index for Loop[iLoop] is equal to Loop[iLoop −1].Last +1. The integer foreground color can be either a color transformed to an integer, if the image is a color one, or an index, if the image is an indexed one.

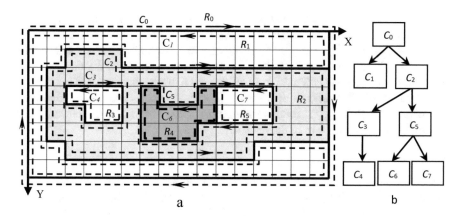

Fig. 4.4   Boundary curves (loops) C_i, $i \in [0, 7]$ (**a**) and the inclusion tree (**b**) of the image of Fig. 4.3

The algorithm "Trace" has always the starting direction 3, i.e., against the positive Y-axis, thus leaving the pixels with the foreground color on the left side of the direction of tracing. "Trace" labels the visited cracks of the direction 3 only since "Search" looks only for these cracks. The labels are saved not in the image but rather in a special array "Lab" which has the same number of rows as the image, whereas the number of columns is by 1 greater than that of the image. This is necessary to obtain memory space for the labels of the cracks at the right-hand border of the image. These cracks are the starting cracks for components touching the right-hand border of the image, as e.g., for the component R_2 in the image of Fig. 4.3 (above) and of Fig. 4.4 (above). The algorithm "Trace" works with standard coordinates. Therefore, cracks are not represented as a variable in "Trace". A crack is present in "Trace" only indirectly as its direction and as a pixel of "Lab" which is the pixel whose standard coordinates are equal to the standard coordinates of the crack.

In the following algorithm "Trace" the notation "Image" stands for the original image. The variable "color" is the foreground color or foreground index of the subset whose boundary (outer or inner) is to be traced; "x" and "y" are the coordinates of the starting point; "dir" is the direction of the running (virtual) crack leading from one point to the next one; "Loop" is the global variable representing the structure containing the coordinates of the starting point of the running loop, the integer foreground color of the incident region and the index "Last" pointing to the last direction of a crack in the array "Directions[]" containing the directions of the boundary cracks of all loops. The arrays "Loop[]" and "Directions[]" are global. "InSpace(C)" is a subroutine which returns TRUE if the cell "C" is inside of the image; otherwise, it returns FALSE. The algorithm "Trace" uses the arrays "toLeft [dir]", "toRight[dir]" and "Step[dir]" of two-dimensional constant vectors specifying correspondingly the displacement to the left from the crack with direction "dir", to the right from the crack with direction "dir" or straight ahead along the crack with direction "dir". The parameter "nLoop" of the algorithm "Trace" defines the number of already detected loops and the parameter "nDir" defines the number of already detected directions of the cracks in the loops. These variables must be saved in such a way that they can be accessible to the algorithm "Restore" described below. The utility algorithm "GetSingular(Point, ..., color)" return the logical value TRUE if the zero-dimensional cell "Point" is singular, which means that the set of pixels with the value "color" bounded by "Point" is disconnected. The utility algorithm "EquNaLi(..., Point, ...)" returns the optimal color of the zero-dimensional cell "Point" as described above in the Sect. 4.1.

Algorithm Trace(Image, x, y, nLoop, nDir, color):

The algorithm "Trace" uses the fields "LeftOf", "RightOf" which are defined as follows:
  LeftOf[0]=(0, -1); LeftOf[1]=(0, 0); LeftOf[2]=(-1, 0); LeftOf[3]=(-1, -1); RightOf[0]=(0, 0);
  RightOf[1]=(-1, 0); RightOf[2]=(-1, -1); RightOf[3]=(0, -1). The field "Step[4]" is described in
  the text of Section 3.1.

Step 1: Declare the variables Left, Right and Point as two-dimensional vectors.
Step 2: Declare the variables dir, Out, leftColor, rightColor, and PointColor.
Step 3: Initialize dir ← 3, Out ← -1, Point.X ← x, Point.Y ← y.
Step 4: Set Loop[nLoop].Start ← Point.
Step 5: Repeat the steps if Point.X is not equal x or Point.Y is not equal y:
    5.1: Set Left ← Point + LeftOf[dir]
    5.2: Set Right ← Point + RightOf[dir]
    5.3: If InSpace(Right) is true, then
                 set rightColor ← ToInteger(Image[Right.X, Right.Y]).
      else set rightColor ← Out.
    5.4: If InSpace(Left) is true, then
                 set leftColor ← ToInteger(Image[Left.X, Left.Y]).
      else set leftColor ← Out.
    5.5: If GetSingular(Point, Image, rightColor, leftColor, color, dir) returns TRUE, then
        Begin
          PointColor ← EquNaLi(Image, Point, rightColor, leftColor).
          If PointColor is equal to color, then set dir ← dir + 1.
          else set dir ← dir + 3.
        End
    5.6: If GetSingular(Point, Image, rightColor, leftColor, color, dir) returns FALSE, then
        Begin
          If leftColor is equal to color, then set dir ← dir + 1.
          else
            If rightColor is not equal to color, then set dir ← dir + 3.
    5.7: If dir is greater than 3, then set dir ← dir - 4.
    5.8: If dir is equal to 3, then set Lab[P.X, P.Y-1] ← 1.   (Labeling)
    5.9: Set Directions[nDir] ← dir.   (Record of the crack code)
    5.10: Set nDir ← nDir + 1.
    5.11: Set Point ← Point + Step[dir].   (a step in the new direction)
    5.12: If Point.X is equal to x and Point.Y is equal to y, then stop repetition step 5.
Step 6: Set Loop[nLoop].Color ← color.
Step 7: Set Loop[nLoop].Last ← nDir - 1.
Stop.

Results of tracing boundaries in the image of Fig. 4.3 (above) and their inclusion tree are shown in Fig. 4.4 (above).

We describe the algorithm "MakeTree" producing the inclusion tree below. Now we shall formulate the most important properties of the loops.

4.2.1   Properties of the Loops

The boundary of a subset S of the image I is the set of all cracks and points whose smallest open neighborhood (SON) crosses both S and its complement $I - S$.

1. If the subset S is a connected set of all pixels with one and the same color or index C, then the boundary of S is a set of closed curves called loops. This is true because from the site of C the image is a binary one: C is foreground and "*not C*" is background. It is well known that each boundary in a binary image is a set of closed curves. Each loop in the boundary of S obtains C as its proper foreground color which we call in what follows the "forecolor".

2. If a loop contains its forecolor in its interior, then it is called a *positive*, otherwise a *negative* loop. The sequence of the directed cracks of a positive loop runs counterclockwise; that of a negative loop runs clockwise. A sign can be assigned to each loop. The sign of a loop can be defined by the direction of its first crack called start direction: If the start direction is 0 (to the right), then the loop is negative; if it is 2 (to the left), then the loop is positive.

3. The boundary of the whole image can also be considered as a loop with the not existing forecolor "Out" or -1. This loop is negative since it does not contain its forecolor.

4. It is possible to calculate an inclusion *tree* for all loops of an image. The loops of an image are ordered: each loop obtains a number according to the order of the calculation. A loop obtains in the inclusion tree three labels: *father, son,* and *brothe*r. The father of the loop L is the number of the loop containing L immediately. "Immediately" means that there is no other loop that lies in the father loop and contains L in its interior.

   The *son* of the loop L is the eldest son of L which means that *son* is the smallest number among the numbers of the loops immediately contained in L.

   Brother of the loop LB is the eldest brother of LB. A brother can be a brother of the son or the brother of another brother. This means that the *brother* of the loop LB is the smallest number among the numbers of the loops immediately contained in the loop "father of LB" and being greater than the number of the loop LB.

5. Each negative loop LN contains several but at least one positive loop. The number of pixels in a loop is the area of the loop. The positive loops contained in the negative loop LN contain all in the loop LN contained pixels so that the sum of the areas of these positive loops is equal to the area of LN.

6. A negative loop LN has no immediately contained pixels. This means that if a pixel is contained in LN, then it is also contained in a positive loop immediately contained in LN.

7. The negative and positive loops of an image form layers in the inclusion tree: the layer 0 is the singe negative loop "Loop[0]" containing the whole image. The layer 1 is formed by all positive loops which are immediately contained in "Loop[0]". The layer 2 is formed by all negative loops which are immediately contained in the positive loops of the layer 1 etc. General: The layer (n + 1) is formed by all positive loops immediately contained in the negative loops of the layer (n). The layer (n + 2) is formed by all negative loops which are immediately contained in the positive loops of the layer (n + 1).

8. A negative loop can be immediately contained only in a positive loop.

9. A positive loop can be immediately contained only in a negative loop.

10. A connected set of pixels of a single color or index can be immediately contained *only* in a positive loop *LP*. The area of such a set is equal to the area of the positive loop *LP* minus the sum of the areas of all the negative loops immediately contained in this positive loop *LP*.

11. If the father of the loop *L* has the same *forcolor* as *L*, then *L* is a negative loop. If, however, the father of the loop *L* has a *forcolor* different from the forcolor of *L*, then *L* is a positive loop.

4.2.2   The Sub-Algorithm "MakeTree"

Let us describe the sub-algorithm "MakeTree".

The inclusion tree describes the relations between the loops indicating which loops are contained in the interior of another loop and which loop contains the considered loop in its interior. The structure of a loop obtains additional members "father", "son", and "brother". The father of a loop *L* is the index of the loop immediately containing the loop *L* in its interior. "Immediately" means that there is no such loop in the interior of the father that contains *L* in its interior. The son of a loop *L* is the smallest index of the loops immediately contained in the interior of *L*. Thus, the son is, so to say, the eldest son of *L*. The father of the son of *L* is *L*. The brother of the son of the loop *L* is the loop with the smallest index of the loops immediately contained in the interior of *L* and different from the son. Thus, brother is the eldest brother of son. The father of the brother is *L*. The brother of a brother *B* is the next eldest brother of *B*.

The algorithm "MakeTree" starts with calculating for each loop found by the algorithm "Search" a box, which is a structure containing four values: MinX, MinY, MaxX, and MaxY. The box is a member of the structure of a loop. The value MinX of the loop *L* is the smallest value of the standard x-coordinate of all points contained in the loop *L*. The remaining three values are similar. Calculating of the box can be performed by the subroutine "MakeBox".

Then the algorithm calculates the parameters of the loop "Loop[0]". We did not calculate the code of this loop in "Search" because the code is exceptionally large, but it can easily be calculated from the size of the image. The part of this loop running at the right-hand border of the image is necessary to calculate the father of the positive loops lying inside of "Loop[0]".

Then the algorithm tests all loops while starting with "Loop[0]". It assigns the value -1 to the variable "Loop[0].Father" which means "no father". For each actual "Loop[i]" it checks all loops with an index less than "i" whether it can be the father of "Loop[i]".

Algorithm "MakeTree"

Step 1: Declare variables SignI, SignF, iL and father.
Step 2: Declare logical variables found and Test.
Step 3: Initialize iL ← 0.
Step 5: Repeat the steps until iL is less than nLoop:
    5.1: Call utility algorithm MakeBoxes(iL).
    5.2: Set iL ← iL + 1.   (End repetition Step 5)
Step 6: Set Loop[0].Father ← -1.
Step 7: Set iL ← 1.
Step 8: Repeat steps until iL is less than nLoop:
    8.1: Set found ← false.
    8.2: Set SignI ← Loop[i].Sign.
    8.3: Set father ← i - 1.
    8.4: Repeat steps until father is greater or equal to 0:
    8.4.1: Set SignF ← Loop[father].Sign.
    8.4.2: If SignI*SignF is less than 0, then
        Begin
        Set Test ← return value of PointInLoop(Loop[iL].Start, father).
        If Test is equal to true, then
          Begin
          Set Loop[iL].Father ← father.
          Call utility algorithm PutSon(iL, father).
          Set found ← true.
          Interrupt the repetition 8.4.
        End   (End of If Test is equal to true)
      End   (End if 8.4.2)
    8.4.3: Set father ← father - 1.   (End repetition 8.4)
    8.4.4: If found equals false, then print message "No father for loop iL".
    8.5: Set iL ← iL + 1.   (End repetition step8)
Stop.

We shall describe now the algorithm of the important logical method "PointInLoop(Point, iLoop)". This method returns the value "true" if "Point" lies inside of the "Loop[iLoop]". It tests first whether "Point" lies inside the box of "Loop[iLoop]". If it is outside the box, then the method returns the value "false". Otherwise, it runs through all cracks of "Loop[iLoop]" and counts the number of cracks lying in the beam from the pixel incident with "Point" to the right border of the image. The method returns "true" if the count is odd, or it returns "false" if the count is even.

The parameter "Point" is a two-dimensional vector presenting the standard coordinates of the tested point. The variables "Crack", "auxPoint" and "Pixel" are also two-dimensional vectors. They represent *combinatorial coordinates* of the cells of the complex representing the image. The operation "Modulo X" means the remainder after division by X.

Algorithm "PointInLoop(Point, iLoop)"

Step 1: Declare variables Crack, auxPoint and Pixel as two-dimensional vectors.
Step 2: Declare the variables First, Count, dir, dirOld, I, iDir, and Length.
Step 3: If Point.X is greater than MaxX of the box of Loop[iLoop], then return false.
Step 4: If Point.X is less than MinX of the box of Loop[iLoop], then return false.
Step 5: If Point.Y is greater than MaxY of the box of Loop[iLoop], then return false.
Step 6: If Point.Y is less than MinY of the box of Loop[iLoop], then return false.
Step 7: Set Pixel.X ← 2*Point.X - 1.
Step 8: Set Pixel.Y ← 2*Point.Y + 1.
Step 9: Set auxPoint.X ← 2*Loop[iLoop].Start.X.
Step 10: Set auxPoint.Y ← 2*Loop[iLoop].Start.Y.
Step 11: If iLoop is equal to 0], then set First ← 0.
       else set First ← Loop[iLoop - 1].Last + 1.
Step 12: Initialize Count ← 0, Length ← Loop[iLoop].Last - First + 1, and I ← 0.
Step 13: Repeat the steps until I is less or equal to Length:
    13.1: Set iDir ← First + I (Modulo Length).
    13.2: Set dir ← Directions[iDir].
    13.3: Set Crack ← auxPoint + step[dir].
    13.4: Set auxPoint ← Crack + step[dir].
    13.5: If Crack.Y equals Pixel.Y and Crack.X is greater than Pixel.X and
         dir is odd, then set Count ← Count +1.
    13.6: Set dirOld ← dir.
    13.7: Set I ← I + 1.   (End step 13.)
Step 14: If Count is odd, then return true.
       else return false.
Stop.

We do not describe the algorithm of the method "MakeBoxes" because it is evident.

Here is the algorithm of the method "PutSon(iSon, iFather)":

The Algorithm "PutSon"(iSon, iFather)

Step 1: Declare variables Count and next.
Step 2: If Loop[iFather].Son is less than 0, then
    Begin
      Set Loop[iFather].Son ← iSon.
      Return.
    End
Step 3: Set next ← Loop[iFather].Son.
Step 4: Repeat the steps if the algorithm is not returned:
    4.1: If Loop[next].Brother is less than 0, then
      Begin
        Set Loop[next].Brother ← iSon.
        Return.
      End
      else set next ← Loop[next].Brother.
    4.2; Set Count ← Count + 1.
    4.3: If Count is greater than nLoop, then
      Begin
       print "Error: PutSon is cycling.".
       Return.
      End
Stop.

Below in Table 4.1 is an example of the code produced by the algorithm "Search" for the image shown in Fig. 4.3.

Table 4.1 Code produced by the algorithms Search and MakeTree for the image shown in Fig. 4.3

```
Loop=0 Sign=-1 Start=(0,0) Color=-1 Father=-1 Son=1 Brother=-1.
Directions:
000000000000000001111111122222222222222222233333333

Loop=1 Sign=1 Start=(16,0) Color=1 Father=0 Son=-1 Brother=2.
Directions:
222222222222222211111111100000000000000000033222212222222222
3233330300010000000000033

Loop=2 Sign=1 Start=(5,1) Color=2 Father=0 Son=3 Brother=-1.
Directions:
22212111101000000000030000333322222222223

Loop=3 Sign=-1 Start=(2,3) Color=2 Father=2 Son=4 Brother=5.
Directions:
0001122323

Loop=4 Sign=1 Start=(5,3) Color=1 Father=3 Son=-1 Brother=-1.
Directions:
2221010033

Loop=5 Sign=-1 Start=(6,3) Color=2 Father=2 Son=6 Brother=-1.
Directions:
0100300001122221222333

Loop=6 Sign=1 Start=(7,3) Color=4 Father=5 Son=-1 Brother=7.
Directions:
2111000303321223

Loop=7 Sign=1 Start=(13,3) Color=1 Father=5 Son=-1 Brother=-1.
Directions:
2221100033
```

4.2.3   The Sub-Algorithm "Restore"

It is possible to restore the original color or indexed image from the code of the loops produced by the algorithm "Search". The reconstructed image corresponds exactly to the original image.

The idea of the reconstruction is rather simple. To test whether a pixel P lies in the interior of a given closed curve (a boundary of a homogeneous region is always a closed digital curve) it is necessary to count the intersections of the curve with a ray from P to any point outside the image: If the number of intersections is odd, then the pixel P lies inside; if it is even, then P lies outside. However, the counting is a non-trivial problem since it is difficult to distinguish between intersection and tangency if the boundary is given as a sequence of pixels (Fig. 4.5a, b).

As one can see, the set of intersections of a sequence of pixels with the ray is exactly one and the same in the cases, when the pixel P lies inside or outside the closed sequence. The solution becomes easy if the boundary is given as one or many sequences of alternating cracks and points in a complex and the "ray" is a sequence of alternating vertical cracks and pixels all lying in one row of the grid (Fig. 4.5c).

A boundary in a complex is a closed sequence of cracks and points (Fig. 4.5c). Intersections with a ray consisting of pixels and vertical cracks are possible only at vertical cracks and the problem of distinguishing between intersections and points of tangency does not occur.

The algorithm "Restore" must have access to the global fields "Loop[]" and "Directions[]". The algorithm obtains a new empty image and fills first its pixels with the value "Empty" which must be one which is never present in an image. Then "Restore" reads the saved code loop by loop, calculates the coordinates of subsequent cracks and labels the crack of direction 1 with the color of the actual loop. Then it scans the image row by row, reads each value different from "Empty" and fills the subsequent pixels with the read value until a new value different from "Empty" occurs.

In the following algorithm "Restore" the notation "Rest" stands for the image to be reconstructed. This image must be a color image with 24 bits per pixel in both cases: whether a color image or an indexed image should be reconstructed. In the

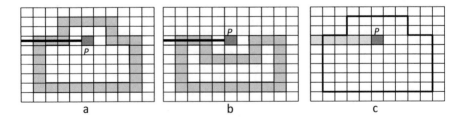

a   b   c

Fig. 4.5 Intersection (**a**) and tangency (**b**) are easy to distinguish in boundaries in a complex (**c**)

latter case the reconstructed indexed image will be converted into a color image by means of the palette produced in the transformation of the original color image into an indexed image or contained in the image if the original image was an indexed one.

As explained above, we transform the channels Red, Green and Blue of a pixel in a color image to an integer variable. This is convenient, since then we can process a color, or an index, or an auxiliary variable "Empty" (see below) in the same way. In the algorithm "Restore" we must make a back transformation from the integer variable representing a color to the three channels Red, Green and Blue. We are using for this purpose the method Shift(iColor, value) which shifts the integer variable "iColor" by "value" bits to the lower valuable bits if "value" is negative, or to the higher bits if "value" is positive.

The value "Empty" is a value which does not occur in the pixels of the original color or indexed image and thus also not among the "for_colors" assigned to the boundary loops. In the algorithm below "Empty" is equal to -1.

The value "Loop[i].Last", as already mentioned above, is the index of the last record in the array "Directions" containing the directions of the cracks of the i-th boundary loop. The value "first" is the index of the first byte of the loop. This value is not saved in the code since it can be easily computed from the saved values of "Last": For the first boundary with $i = 0$ "first" is equal to 0; for any other value of i it is equal to the value of "Last" for the preceding boundary plus 1.

The auxiliary field "Label" having the size of the image "Rest" is filled by the value "Empty". Then the repetition over all loops is started. The variable "Pixel" is set equal to the starting point of the actual loop. Then the directions "dir" are read from the field "Directions". If the actual direction is equal to 1, then Label[Pixel] is set equal to the color of the actual loop. The variable "Pixel" is moved by one step into the direction "dir".

The part following the repetition over the loops performs the filling of the pixels of the image "Rest". This part moves the variable "Pixel" through the field "Label". If Label[Pixel] is not equal to "Empty", then the variable "lab" is set equal to Label [Pixel]. Then all elements Rest[Pixel] are set equal to "lab". Note that in all three algorithms of this Chapter only the standard coordinates are employed. This is possible because, according to the Coordinate Assignment Rule described in [2], Sect. 7.2.1, p. 118, the standard coordinates of a point, of a crack and of a pixel "owing" that point and crack have the same values.

Algorithm Restore(Rest, Palette)

The variable 'nLoop' and the fields 'Loop[nLoop]' and 'Directions are global.

The expression "Shift(Palette[lab], -8*ic)" denotes the integer value "Palette[lab]" shifted in the direction of lower valuable bits by 8*ic bits.

Step 1: Declare the variable Pixel as a two-dimensional vector.
Step 2: Declare the variables ic, Empty, first, iDir, lab, and maxColor.
Step 3: Initialize the variables Empty ← -1 and maxColor ← 0.
Step 4: Declare two-dimensional array Label of the size Rest.width*Rest.height.
Step 5: Initialize all elements of Label as equal to Empty.
Step 6: Initialize iL ← 1.
Step 7: Repeat until iL is less than nLoop:
    7.1: Set Pixel ← Loop[iL].Start.
    7.2: Set first ← Loop[iL -1].Last + 1.
    7.3: Set iDir ← first.
    7.4: Repeat until iDir is less or equal to Loop[iL].Last:
    7.4.1: If Directions[iDir] is equal to 1, then
        Begin
          Set Label[Pixel.X, Pixel.Y] ← Loop[iL].Color.
          If Loop[iL].Color is greater than maxColor, then
                  set maxColor ← Loop[iL].Color.
          Set Pixel.Y ← Pixel.Y + 1.
        End
      else
        If Directions[iDir] is equal to 0, then set Pixel.X ← Pixel.X + 1.
        else
          If Directions[iDir] is equal to 2, then set Pixel.X ← Pixel.X - 1.
          else
            If Directions[iDir] is equal to 3, then set Pixel.Y ← Pixel.Y - 1.
    7.4.2: Set iDir ← iDir + 1.   (End repetition 7.4.)
    7.5: Set iL ← iL + 1.   (End repetition 7.)
Step 8: Set Pixel.Y ← 0.
Step 9: Repeat until Pixel.Y is less than Rest.height:
    9.1: Set Pixel.X ← 0.
    9.2: Repeat until Pixel.X is less than Rest.width:
    9.2.1: If Label[Pixel.X, Pixel.Y] is not equal Empty, then
               set lab ← Label[Pixel.X, Pixel.Y].
    9.2.2: If maxColor is greater than 255, then
        Begin 9.2.2.1
          9.2.2.1: Set ic ← 0.
          9.2.2.2: Repeat until ic is less than 3:
          Begin 9.2.2.2
          If ic is equal to 0, then
             set Rest[ic, Pixel.X, Pixel.Y] ← lowest 8 bits of lab.
          else set Rest[ic, Pixel.X, Pixel.Y] ← lowest 8 bits of Shift(lab, -8*ic).
          Set ic ← ic + 1.
          End 9.2.2.2   (End of Repeat until ic is less than 3.)
        End 9.2.2.1   (End of the first part of If maxColor is greater …)
      else   (maxColor is less than or equal to 255.)
        Begin 9.2.2.3
        Set ic ← 0.
        Repeat until ic is less than 3:
          Begin 9.2.2.4
          Set Rest[ic, Pixel.X, Pixel.Y] ←
                  lowest 8 bits of Shift(Palette[lab], -8*ic).
          Set ic ← ic + 1.
          End 9.2.2.4   (End of Repeat until ic is less than 3.)
        End 9.2.2.3   (End of "else" and of If maxColor is greater …)
    9.2.3: Set Pixel.X ← Pixel.X + 1.   (End of repetition 9.2.)
    9.3: Set Pixel.Y ← Pixel.Y + 1. .   (End of repetition 9.)
  Stop.

Fig. 4.6 Example of an image indexed with 183 colors (**a**) and the loops with at least 300 cracks found by the algorithm "Search" (**b**)

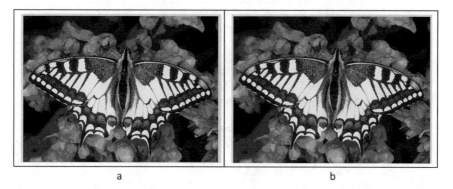

Fig. 4.7 Example of an image indexed with 219 colors (**a**) and the image restored from the loops found by the algorithm "Search" (**b**)

Let us demonstrate some examples of detecting boundaries of regions with constant indexes in an indexed image (Fig. 4.6).

Here is one more example presenting the result of restoring the image from the loops. The restored image and the original indexed image are identical (Fig. 4.7) because the restoration is precise.

4.2.4 Tracing Equations of Curves

As already mentioned in [2], p. 199, it is possible to draw a curve defined not as the boundary of a subset of pixels in an image but defined by means of an equation, as e.g. $F(x, y) = 0$. The value $F(x, y)$ is either negative, or non-negative. There are only two possibilities. Therefore, the equation can be traced in the same way as a boundary in a binary image. The pixels in which the left side of the equation is positive or equal to 0 compose the subset, and the negative pixels compose the complement. This way of tracing curves is important if the equation is not solvable to be represented as $y = f(x)$. This is e.g., the case for the cardioids with the equation

$$\left(x^2 + y^2\right)^2 - 2 * a * x * \left(x^2 + y^2\right) - a^2 y^2 = 0.$$

Here is the algorithm:

Algorithm TraceEquation( ):

The fields "toLeft4]", "toRight[4]", and "Step[4]" are described in Section 3.1.

Step 1: Start.
Step 2: Declare variables a, dir, valLeft, valRight, x0, and y0.
Step 3: Initialize variables: a ← 100; x0 ← a / 2; y0 ← 3*a/2; dir ← 1.
Step 4: Declare two-dimensional vectors Crack, Point, StartP, Left and Right
Step 5: Initialize variables StartP.X ← 2*x0; StartP.Y ← 2*y0 and Point ← StartP.
Step 6: Repeat the steps: (Stop condition in step 6.10)
    6.1: Set Left ← Point + toLeft[dir].
    6.2: Set Right ← Point + toRight[dir].
    6.3: Set valLeft ← Equation(Left.X/2, Left.Y/2, a).
    6.4: Set valRight ← Equation(Right.X/2, Right.Y/2, a).
    6.5: If valRight is negative, then set dir ← dir + 1.
      else
        if valLeft is positive or equal to 0, then set dir ← dir + 3.
    6.6: If dir is greater than 4, then set dir ← dir - 4.
    6.7: Set Crack ← Point + Step[dir].
    6.8: If dir equals 1 or dir equals 3, then
        draw a line from (Crack.X, Crack.Y-1) to (Crack.X, Crack.Y+1).
      else
        draw a line from (Crack.X-1, Crack.Y) to (Crack.X+1, Crack.Y).
    6.9: Set Point ← Crack + Step[dir].
    6.10: If Point is equal to StartP, then break the repetition. (End repetition 6)
Stop.

An example of the curve drawn by the above algorithm is shown in Fig. 4.8.

4.3 Encoding Boundaries with the Algorithm "MakeLineList"

A drawback of the CORB algorithm is that some parts of the boundaries are traced and encoded twice. We describe in this section the main algorithm "MakeLineList" which is free of this drawback; however, it is more complicated than the CORB algorithm. This algorithm is a simplified version of the Cell-List-Algorithm developed by the author [1] and shortly described in [2], pp. 212–217. We consider here only the version usable for indexed images because, as already mentioned above, color images mostly contain an exceptionally large number of different colors and therefore homogeneous regions, whose boundaries are processed, are in a color image exceedingly small. Many homogeneous regions consist of a single pixel.

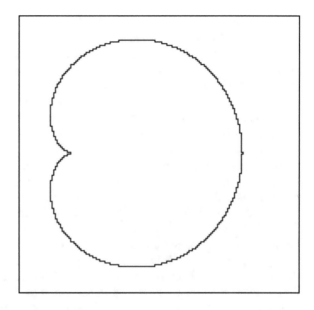

Fig. 4.8 Cardioid with a = 50 drawn by the algorithm "TraceEquation"

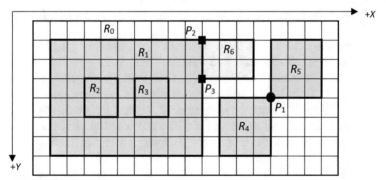

Fig. 4.9 Examples of blocks and regions

Consider now an indexed image I and the set of all pixels of I which have a fixed index, e.g., of the set $R = R_1 \cup R_4 \cup R_5$ in Fig. 4.9 above. This is the disconnected set of all dark gray pixels. The components of the set R are the sets R_1, R_4 and R_5 if the point P_1 does not belong to R. We regard these components in connection with the Cell-List-Algorithm as *regions of a constant color* instead of 2-blocks defined in [2]. In contrast to a block, a region does not have to be homeomorphic to an open ball. A region can have some holes and some singular points.

Consider the example of Fig. 4.9.

The set R_1 of dark gray pixels without its boundary is a region. A point incident to at least 3 pixels with different indices is regarded as a *branch point* (BP). For example, the points P_2 and P_3 are branch points, whereas the point P_1 is not since there are only two indices in its neighborhood. The point P_1 is important for describing the boundaries of the regions R_4 and R_5. It is the starting point of the sequences of cracks composing the boundaries of R_4 and R_5. These boundaries contain no branch points; they are simple closed curves. We call both the branch points as well as starting points of closed curves the *important points* (IP).

A connected subset of the boundary of a region lying between two BPs is regarded as a *line*. If the boundary of a region contains no branch points, then the boundary without one point (the starting point of the tracing) is also regarded as a line. The starting point is an important point (IP).

We consider here only the case in which the image is represented as a two-dimensional array of pixels. Thus, the image can be regarded as a two-dimensional Cartesian complex, where each 0-cell is incident to at most four cracks. Therefore, each IP is incident to at most four lines.

We describe here the structure "LineList" (a class in the program language C#) to be developed. It contains the following three fields: the field "Points" containing the important points IP, the field "Lines" containing the lines, and the field containing the metric data. The latter field may be the field of bytes if the lines are encoded by the directions of subsequent oriented cracks; or it is the field of indices of the starting points of the digital straight segments (compare the description in Sect. 5.1), or of the vertices of the polygons approximating the lines (compare the description in Sect. 5.4).

The structure "LineList" also contains the auxiliary fields "InStack", "TracedCrack", and "Complex". The field "InStack" is an array of integers. The value of InStack[P] greater than 0 means that the point P was already pushed to the stack. The field "TracedCrack" serves for labeling the locations already visited by the algorithm "TraceC". Thus, the algorithm "MakeLineList" obtains the possibility to skip these locations and to prevent a multiple processing of one and the same component of the boundaries. The field "Complex" serves for labeling the Cracks belonging to some boundary.

The fields "Points" and "Lines" are arrays of structures. The structure of an element of "Points" contains a two-dimensional vector "Point" containing two integer coordinates of the point. It also contains four values corresponding to four directions 0, 1, 2 and 3 of the oriented cracks incident to the point. If a crack is a boundary crack but the index of the line containing this crack is jet unknown, then the value in "Points" corresponding to the crack is equal to -1. If, however, the crack is no boundary crack, then the corresponding value is equal to -2. We do not employ here signs of the indices of lines to indicate the orientation of a line, i.e., to show whether the line goes away from the point, or it comes to the point. The orientation of a line is defined by its starting and end points.

The structure of an element of the field "Lines", which is an oriented line, contains a two-dimensional vector of the starting point of the line, the color index "indexL" (the image is an indexed one) of the homogeneous region to the left-hand side of the line, the index "indexR" of the region to the right-hand side and the

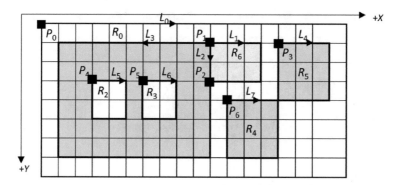

Fig. 4.10 Artificial test image with indicated important points and lines

indices "first" and "last" of the beginning and the end of the metrical data belonging to the line.

The structure "LineList" contains besides the mentioned fields the variables "nPoint", "nLine" and "nByte" indicating the numbers of already detected points, lines, and metrical data.

Related to the structure "LineList" are the algorithms "MakeLineList", "Component", "TraceC" and "RestoreC".

All the mentioned fields and variables are directly accessible to these four algorithms related to the structure "LineList", and therefore they are not defined in the algorithms.

Figure 4.10 above shows the artificial test image used to explain the contents of the structure "LineList". Note that the positive Y-axis in Fig. 4.10 is directed downwards.

The Table 4.2 shows the line list of the artificial image presented in Fig. 4.10. The indices of points, lines and bytes correspond exactly to that of Fig. 4.10.

We describe below the main algorithm "MakeLineList". It scans the indexed image row by row, finds two horizontally adjacent pixels with different indices "IndexOld" and "Index". If the found location is yet not labeled as already visited by the sub-algorithm "TraceC", then the algorithm "Component" is called with the given image, and the two-dimensional vector "StartPoint" corresponding to the coordinates of the pixel with the index "Index", transmitted as arguments.

The sub-algorithms "Component", "TraceC" and "RestoreC" are described below. The notations in the following algorithm "MakeLineList" are the same as in the above-mentioned version of the CORB algorithm (Sect. 4.2). The variable "nByte" is global. The global two-dimensional field "Label" has the size of the Image.

Table 4.2 The contents of the structure "LineList" for the artificial image of Fig. 4.10

List of lines:
Line=0; Start=(0, 0); indL=-1; indR=1; First byte=0; Last byte=51.
Line=1; Start=(10, 1); indL=1; indR=2; First byte=52; Last byte=59.
Line=2; Start=(10, 1); indL=2; indR=3; First byte=60; Last byte=61.
Line=3; Start=(10, 1); indL=3; indR=1; First byte=62; Last byte=89.
Line=4; Start=(14, 1); indL=1; indR=3; First byte=90; Last byte=101.
Line=5; Start=(3, 3); indL=3; indR=1; First byte=102; Last byte=109.
Line=6; Start=(6, 3); indL=3; indR=1; First byte=110; Last byte=117.
Line=7; Start=(11, 4); indL=1; indR=3; First byte=118; Last byte=129.

List of points:
Point=0; Loc=(0, 0); IncLine=(0; 0; -2; -2)
Point=1; Loc=(10, 1); IncLine=(1; 2; 3; -2)
Point=2; Loc=(10, 3); IncLine=(1; 3; -2; 2)
Point=3; Loc=(14, 1); IncLine=(4; 4; -2; -2)
Point=4; Loc=(3, 3); IncLine=(5; 5; -2; -2)
Point=5; Loc=(6, 3); IncLine=(6; 6; -2; -2)
Point=6; Loc=(11, 4); IncLine=(7; 7; -2; -2)

List of bytes:
Line=0 bytes: 0, 0, 0, 0, 0, 0, 0, 0, 0, 0, 0, 0, 0, 0, 0, 0, 0, 0,
1, 1, 1, 1, 1, 1, 1, 2, 2, 2, 2, 2, 2, 2, 2, 2, 2,
2, 2, 2, 2, 2, 2, 2, 3, 3, 3, 3, 3, 3, 3, 3,
Line=1 bytes: 0, 0, 0, 1, 1, 2, 2, 2,
Line=2 bytes: 1, 1,
Line=3 bytes: 2, 2, 2, 2, 2, 2, 2, 2, 2, 1, 1, 1, 1, 1, 1, 0, 0, 0,
0, 0, 0, 0, 0, 0, 3, 3, 3, 3,
Line=4 bytes: 0, 0, 0, 1, 1, 1, 2, 2, 2, 3, 3, 3,
Line=5 bytes: 0, 0, 1, 1, 2, 2, 3, 3,
Line=6 bytes: 0, 0, 1, 1, 2, 2, 3, 3,
Line=7 bytes: 0, 0, 0, 1, 1, 1, 2, 2, 2, 3, 3, 3,

Algorithm MakeLineList(Image[width, height])

Step 1: Declare the integer variables Index, lab, IndexOld, x, and y
and the two-dimensional vector StartP.
Step 2: Initialize all elements of Label ← 0.
Step 3: Initialize y ← 0.
Step 4: Repeat until y is less than height:
  Step 4.1: Set IndexOld ← -1 and x ← 0.
  Step 4.2: Repeat until x is less than width:
      4.2.1: Set Index ← Image[x, y].
      4.2.2: Set lab ← Label[x, y].
      4.2.3: If Index is not equal IndexOld and lab is not equal 0, then
          Begin
            Set StartP ← vector (x, y).
            Call algorithm Component (Image, StartP).
          End
      4.2.4: Set IndexOld ← Index.
      4.2.5: Set x ← x + 1.   (End of the repetition of step 4.2).
  Step 4.3: Set y ← y + 1.   (End of the repetition of step 4).
  Stop

The sub-algorithm "Component" traces and encodes one connected component of the set of all boundary elements of the image. This is the component containing the starting point "StartP(x, y)". The algorithm "Component" saves in the fields of the structure "LineList" all data describing the component.

The algorithm "Component" operates as follows. It starts the sub-algorithm "TraceC" which traces the boundary beginning at the starting point "StartP" and running until a branch point or until the starting point. In the latter case the traced boundary contains no branch points. The component consists in this case only from the starting point and a closed line beginning and ending at the starting point. The algorithm "Component" saves the coordinates of the starting point and the data describing the closed line in the structure "LineList" and stops.

If, however, a branch point was found, then "Component" discards the data of the tracing, puts the coordinates of the found branch point into the stack and starts a repetition through all lines and all branch points of the connected component of the boundaries to which the found branch point belongs.

During the repetition, a point P is popped from the stack and all four cracks incident to it are tested for the possibility to start a tracing of the boundary beginning with that crack. Each point has its structure "IncLine" consisting of four values corresponding to the four cracks incident with the point. Each value indicates whether the corresponding crack is a boundary one or the value is equal to the index of an already detected line containing this crack. If the crack is a boundary crack and there is yet no detected line containing this crack, then the tracing starting with this crack until a next branch point is performed. This tracing runs along a line starting and ending with a branch point. The data describing the line are saved in the "LineList" and the next crack incident to the point P is tested.

When all cracks incident to the point P are tested, the next point is popped form the stack and so on until the stack is empty.

It is possible that during the tracing of the boundary a new branch point is discovered. Their coordinates are entered into the "LineList", pushed into the stack and the total number of the branch points is increased. Due to this action the duration of the repetition through all branch points of the component becomes prolonged. The process stops when the stack is empty. The encoding of the component is then finished, and the algorithm "Component" stops. Then the algorithm "MakeLineList" looks for the next component until all component of the image are processed. Now we describe the algorithm "Component" for indexed images. "Component" obtains as the first argument the "Image[width, height]" with the indices of the indexed image. The variables "nByte" and "nLine" are global. "Mod4(x)" is the remainder of dividing x by 4.

The sub-algorithm "TraceC" should be defined in such way that the values of its arguments can be changed inside of "TraceC". Remain that the algorithm "Component" has a direct access to the fields "Points", "Lines", "InStack", and to the variables "nPoint" and "nLines".

Algorithm Component(Image[width, height], vector StartP)

Step 0: Define and initialize integer variables CNX ← 2 * width + 1, CNY ← 2 * NY + 1.

Step 1: Define and initialize the field step[4] of vectors:

step[0] ← (1, 0); step[1] ← (0, 1); step[2] ← (-1, 0); step[3] ← (0, -1).

Step 2: Define integer variables aPop, aPterm, dir, dirBefore, Dir, and il.

Step 3: Define the vectors P, Pcomb, Ppop, Pterm, and Crack.

Step 4: Initialize Line[nLine].first ← nByte and Pterm ← StartP.

Step 5: Call the algorithm TraceC(I, Pterm, dir).

  Begin 5

    5.1: Set rvTrace ← result of TraceC.

    5.2: Set aPterm ← nPoint.

  End 5

Step 6: If rvTrace is equal to 2, then   (Trace has found a closed line)

  Begin 6

    6.1: Set Line[nLine].StartP ← Pterm.

    6.2: If Pterm.Y is positive, then

        set Line[nLine].indexL ← Image[Pterm.X, Pterm.Y - 1].

      else set Line[nLine].indexL ← -1.

    6.3: Set Line[nLine].indexR ← Image [Pterm.X, Pterm.Y].

    6.4: Set Line[nLine].last ← nByte - 1.

    6.5: Call utility algorithm SetIncLines(I, Pterm, aPterm).   (explained below)

    6.6 Set ImPoint[aPterm].Point ← Pterm.

    6.7 Set ImPoint[aPterm].IncLine[0] ← nLine.

    6.8 Set ImPoint[aPterm].IncLine[Oposit(dir)] ← nLine.   (Oposit(dir) is Mod4(dir+2).)

    6.9 Set nLinen ← nLine + 1.

    6.10 Set nPoint ← nPoint + 1.

  End 6

Step 7: If rvTrace is equal to 3, then   (Trace has found a branch point)

  Begin 7

    7.1: Set nByte ← nByte0.

    7.2: Push Pterm to Stack.

    7.3: Repeat until the Stack is not empty: (do-while loop)

      Begin 7.3

        7.3.1: Pop the point Ppop from Stack.

        7.3.2: If InStack[Ppop] is equal to 0, then set InStack[Ppop] ← 1.

        7.3.3: If Ppop is inside the image, then define the values of aPop by means of

            utility algorithm: aPop ← GetAddress(Ppop, nPoint).

          else set aPop ← nPoint.

        7.3.4: If InStack[Ppop] equals 0, then

          Begin 7.3.4

          Set nPoint ← nPoint + 1.

          Call the utility algorithm 'SetIncLines(I, P, aPop)'.

          Set Points[aPop] Ppop.

          End 7.3.4

        7.3.5: Set Pcomb.X ← 2*P.X.

        7.3.6: Set Pcomb.Y ← 2*P.Y.

        7.3.7: Set Dir ← 0.

7.3.8: Repeat until Dir is less than 4:   (The for (Dir…) loop ===)
      Begin 7.3.8
        7.3.8.1: Set Crack ← Pcomb + step[Dir].
        7.3.8.2: If Crack is in image and Points[aPop].IncLine[Dir] equals -1, then
        Begin 7.3.8.2
          7.3.8.2.1: If InStack[Ppop] equals 0 set InStack[Ppop] ← 1.
          7.3.8.2.2: Set Pterm ← Ppop.
          7.3.8.2.3: Set Line[nLine].first ← nByte.
          7.3.8.2.4: Set Line[nLine].StartP ← Ppop.
          7.3.8.2.5: Set PixL ← Crack + step[Mod4(Dir + 3)].
          7.3.8.2.6: Set PixR ← Crack + step[Mod4(Dir + 1)].
          7.3.8.2.7: If PixL is in image, then set Line[nLine].indexL ← I.[PixL/2].
                  else set Line[nLine].indexL ← -1.
          7.3.8.2.8: If PixR is in image, then set Line[nLine].indexR ← I.[PixR/2].
                  else set Line[nLine].indexR ← -1.
          7.3.8.2.9: Set Points[aPop].IncLine[Dir] ← nLine.
          7.3.8.2.10: Set dirBefore ← Dir and dir ← Dir.
          7.3.8.2.11: Call algorithm TraceC(I, Pterm, dir).
          7.3.8.2.12: Set Lines[nLine].last ← nByte – 1.
          7.3.8.2.13: If InStack[Pterm] equals 0, then   (Pterm is a new point)
                Begin 7.3.8.2.13
                    Set aPterm ← nPoint.
                    Call utility algorithm SetIncLines(Image, Pterm, aPterm).
                    Set Points[aPterm].Point ← Pterm.
                    Push Pterm.to Stack.
                    Set nPoint ← nPoint + 1.
                End 7.3.8.2.13
                else set aPterm ← GetAddress(Pterm, nPoint).
          7.3.8.2.14: Set Points[aPterm].IncLine[Mod4(dir+2)] ← nLine.
          7.3.8.2.15: If InStack[Pterm] is equal to 0, then set InStack[Pterm] ← 1.
          7.3.8.2.16: Set nLine ← nLine + 1.
        End (7.3.8.2 If Crack is in image …)
        Step 7.3.7.3 Set Dir ← Dir + 1.
      End 7.3.8   (End Repeat 7.3.8, until Dir is less than 4)
     End 7.3   (End Repeat 7.3, the do-while loop)
    End 7   (End of step 7, If rvTrace is equal to 3.)
Stop

The sub-algorithm "Component" uses the utility algorithms "SetIncLines" and "GetAddress". Here are the presentations of these algorithms.

Utility Algorithm SetIncLines(Image[width, height], vector P, int iPoint)

Step 1: Define the integer variables Dir and I.
Step 2: Define the integer field Index[] of the size 4.
Step 3: Define and initialize the vector field Pixel[4]:
　　　Pixel[0] ← vector (P.X, P.Y), Pixel[1] ← vector (P.X-1, P.Y),
　　　Pixel[2] ← vector(P.X-1, P.Y-1), Pixel[3] ← vector (P.X, P.Y-1).
Step 4: Set I ← 0.
Step 5: Repeat until I is less than 4:
　　　5.1: If Pixel[I] is in Image, then set Index[I] ← Image[Pixel[I].X, Pixel[I].Y].
　　　　　else set Index[I] ← -1.
　　　5.2: Set I ← I + 1:
Step 6: Set Dir ← 0.
Step 7: Repeat until Dir is less than 4.   (Loop for (Dir …)
　　　7.1: If Dir is equal to 0, then
　　　　　iF Index[3] is not equal to Index[0], then set Points[iPoint].IncLine[0] ← -1.
　　　　　else set Points[iPoint].IncLine[0] ← -2.
　　　7.2: If Dir is equal to 1, then
　　　　　if Index[1] is not equal to Index[0], then set Points[iPoint].IncLine[1] ← -1.
　　　　　else set Points[iPoint].IncLine[1] ← -2.
　　　7.3: If Dir is equal to 2, then
　　　　　if Index[2 is not equal to Index[1], then set Points[iPoint].IncLine[2] ← -1.
　　　　　else set Points[iPoint].IncLine[2] ← -2.
　　　7.4: If Dir is equal to 3, then
　　　　　if Index[3] is not equal to Index[2], then set Points[iPoint].IncLine[3] ← -1.
　　　　　else set Points[iPoint].IncLine[3] ← -2.
　　　7.5: Set Dir ← Dir + 1.   (End Repeat Step 7)
Stop.

Here is the utility algorithm "GetAddress":

Utility Algorithm GetAddress(vector P, int nPoint)

Step 1: Define and initialize the variable I ← 0.
Step 2: Repeat until I is less or equal to nPoint:
　　　Begin
　　　If Points[I].Point is equal to P, then set the result ← I and break the repetition.
　　　Set I ← I + 1.
　　　End
Stop.

We present below the algorithm "TraceC".

Algorithm TraceC(Image I[width, height], vector Pterm, int dir)

Step 1: Define and initialize the integer variable i ← 0.
Step 2: Define integer variables StartIndexL, StartIndexR, IndexLeft, and IndexRight.
Step 3: Define the two-dimensional vectors P, Pcomb, Pterm, Ctest, PixLeft,
                                        PixRight, P, Pold, Pstop, and Crack.
Step 4: Define logical variables Branch and AtStart.
Step 5: Define and initialize the vector field step[4]: step[0] ← vector(1, 0),
                    step[1]←vector(0, 1), step[2]←vector(-1, 0), step[3]←vector(0, -1).
Step 6: Define and initialize the vector field Norm[4]: Norm[0] ← vector(0, 1),
                Norm[1]←vector(-1, 0), Norm[2]←vector(0, -1), Norm[3]←vector(1, 0).
Step 7: Set P ← 2*Pterm.   (Version with combinatorial coordinates)
Step 8: Set Pstop ← P.
Step 9: Set Crack ← P + step[dir].
Step 10: Set PixLeft ← Crack – Norm[dir].
Step 11: Set PixRight ← Crack + Norm[dir].
Step 12: Set StartIndexL ← I[PixLeft.X, PixLeft.Y].
Step 13: Set StartIndexR ← I[PixRight, PixRight.Y].
Step 14: If dir is equal to 3 or to 1 and Crack.X is less than I.width/2, then
            set TracedCrack[Crack/2] ← 1.   (Labeling Crack as already traced)
Step 15: Repeat until P is not a branch point and is not equal to Pstop:
     Begin 15
        15.1: Set Ctest ← P + step[dir].
        15.2: Set PixLeft = Ctest - Norm[dir].
        15.3: Set PixRight = Ctest + Norm[dir].   (Two pixels incident to Ctest)
        15.4: Set IndexLeft ← I[PixLeft.X, PixLeft.Y].
        15.5: Set IndexRight ← I[PixRight.X, PixRight.Y].
        15.6: If IndexLeft is different from StartIndexL and from StartIndexR or
                IndexRight is different from StartIndexL and from StartIndexR, then
                Begin 15.6
                Set Branch ← true.
                Set result of TraceC ← 3.
                Break the repetition Step 15.
                End 15.6
        15.7: If IndexRight is equal to StartIndexL, then
                Begin 15.7
                Set dir ← Mod4(dir + 1).
                Set Crack ← P + step[dir].
                End 15.7
        15.8: If IndexLeftt is equal to StartIndexR and
                IndexRight is not equal to StartIndexL, then
        Begin 15.8
        Set dir ← Mod4(dir + 3).
        Set Crack ← P + step[dir].
        End 15.8
        else set Crack ← Ctest.
        15.9: If dir is equal to 3 or to 1 and Crack.X is less than 2*width, then
                set TracedCrack[Crack] ← 1.

15.10: Set P ← Crack + step[dir].
    15.11: Set Byt[nByte] ← dir.
    15.12: Set nByte ← nByte + 1.
    15.13: If P is equal to Pstop, then set AtStart ← true.
    15.14: If Branch or AtStart is true, then
        Begin 15.14
         If AtStart is true, then
          Begin
           Set result of TraceC ← 2.
           Break the repetition Step 15.
          End
        End 15.14
    15.15: Set Pold ← P.
    End 15   (End of the repetition Step 15)
Step 16: Set Pterm.X ← P.X/2.
Step 17: Set Pterm.Y ← P.Y/2.
Step 18: Return result of TraceC.
Stop

4.3.1   *Restoration of Encoded Images*

The restoration of an encoded mage is important because it gives the possibility to check the correctness of the encoding. The encoding is correct if the reconstructed image exactly corresponds the original one.

We describe below the reconstruction algorithm used for reconstructing indexed images encoded by the algorithm "MakeLineList" described above. This reconstruction algorithm is slightly different from that used for the reconstruction of images encoded by the CORB algorithm described above in Sect. 4.2.3 because the structure of the coded lines is different from the structure of the loops of the CORB algorithm: A code of a line contains two colors or two indexes of colors on the sides of the line while a code of a loop contains only one color or index of a color.

The new restoration algorithm works as follows. It obtains as parameter the image "Rest" for the results of the restoration. This image is filled with value EMPTY which is different from all possible values contained in the original image. For example, if the original image is an indexed one, then it contains values from 0 to 255. Then EMPTY can be defined as -1.

The algorithm defines the utility field "Help" of the size of the image to be restored and sets all elements of "Help" equal to EMPTY. It is necessary to use this utility field because the image "Rest" is an indexed image containing values from 0 to 255. It cannot contain the value EMPTY equal to -1. The field "Help" is defined as a field of a type which can contain both the value -1 and the values from 0 to 255.

The algorithm reads the field "Lines", sets the standard coordinates of the vector "Point" equal to that of the starting point of the actual line and reads the field "Bytes" belonging to the line. It sets the variable "Direction" equal to the value of the actual value of "Bytes", sets both the vectors "PixL" and "PixR" equal to the vector "Point". Then it changes the coordinates of "PixL", "PixR" and "Point" depending on the value of "Direction". If the value of "Direction" is equal to 1 (the actual crack is directed downwards) and the vector "PixL" lies inside of the image, then "Help [PixL]" is set to the value of the left index assigned to the line. Note, that the vector "PixL" can lie outside of the image, at its right-hand side, if the actual line runs at that side. Similarly, if the value of "Direction" is equal to 3 (the actual crack is directed upwards) and the vector "PixR" lies inside of the image, then "Help[PixR]" is set to the value of the right index assigned to the line.

Thus, the labeling of some pixels of the field "Help", lying at the sides of the lines is finished. However, most of the pixels of "Help" remain empty. Now the filling of the field "Help" is started. The algorithm scans the field "Help" line by line and checks each pixel "Help[x, y]" of the field "Help": If "Help[x, y]" is not equal to EMPTY, then the variable "Index" gets the value "Help[x, y]". If, however, "Help[x, y]" is equal EMPTY, then "Help[x, y]" gets the value "Index".

At the end of this scanning the field "Help" contains the restored image. It remains a copying of the field "Help" to the image "Rest".

Here is the algorithm "RestoreC":

Algorithm RestoreC(Image Rest, int nPoint)

Step 1: Define the integer variables Dir, iB, iL, Index, x and y.
Step 2: Define and initialize the variable EMPTY ← –1.
Step 3: Define the two-dimensional vectors P, PixL and PixR.
Step 4: Define the field Help of the size Rest.width*Rest.height.
Step 5: Initialize the whole field Help with EMPTY.
Step 6: Set Dir ← 0.
Step 7: Set iL ← 0.
Step 8: Repeat until iL is less than nPoint:
    8.1: Set P ← Lines[iL].Startp and iB ← Lines[iL].first.
    8.2: Repeat until iB is less or equal to Lines[iL].last:
    8.2.1: Set Dir ← Byte[iB].   (Direction of the crack iB)
    8.2.2: Set PixL ← P and PixR ← P.
    8.2.3: If Dir equals 0, then reduce PixL.Y by 1 and increase P.X by 1.
    8.2.4: If Dir equals 1, then reduce PixL.X by 1 and increase P.Y by 1.
    8.2.5: If Dir equals 2, then reduce PixL.X, PixR.X, PixR.Y and P.X by 1.
    8.2.6: If Dir equals 3, then reduce PixL.X, PixL.Y, PixR.Y and P.Y by 1.
    8.2.7: If Dir equals 1 and PixL lies in the Image, then
        set Help[PixL] ← Line[iL].IndexL.
    8.2.8: If Dir equals 3 and PixR lies in the Image, then
        set Help[PixR] ← Line[iL].IndexR.
    8.2.9: Set iB ← iB + 1.   (End repetition 8.2)
    8.3: Set iL ← iL + 1.   (End repetition 8)
Step 9: Set Index ← 0.
Step 10: Set y ← 0.
Step 11: Repeat until y is less than Rest.height:
    11.1: Set x ← 0.
    11.2: Repeat until x is less than Rest.width:
    11.2.1: If Help[x, y] is not equal to EMPTY, then set Index ← Help[x, y].
        else set Help[x, y] ← Index.
    11.2.2: Set x ← x + 1.   (End repetition 11.2)
    11.3: Set y ← y + 1.   (End repetition 11)
Step 12: Copy the field Help to the image Rest,
Stop

Figure 4.11 shows an example of an indexed image with 240 colors of the size of 295 × 250 pixels (73,750 bytes) which was encoded by the algorithm "MakeLineList" described above and exactly reconstructed by the algorithm "RestoreC". The line list contains 8468 important points, 13,318 lines and 39,196 bytes of direction codes.

Fig. 4.11 Example of an indexed image encoded by the algorithm "MakeLineList" and restored by the "RestoreC" algorithm

References

1. Kovalevsky V. Finite topology as applied to image analysis. Comp Vis Graph Image Proc. 1989;45(2):141–61.
2. Kovalevsky V. Geometry of locally finite spaces. Berlin: Editing house Dr. Bärbel Kovalevski; 2008. ISBN 978-3-9812252-0-4

Chapter 5
Boundary Polygonization

Abstract This chapter describes two methods of encoding a boundary as a sequence of straight-line segments: the first method encodes a boundary as a sequence of short straight-line segments being digital images of analog straight-line segments. The second method calculates for each boundary component a polygon approximating the boundary with an assigned precision. Both methods make a geometrical analysis of the boundary possible.

Keywords Digital Straight Segments (DSS) · Digital half space · Digital half-plane · Coordinate assignment rule · Bases of a DSS · Separating form · Singular and main directions · Recognition of DSS · Additional parameters · Applications of DSS · Length of digital curves · Polygonal approximation · Sector method · Applications of polygonal approximation · Recognition of circles and ellipses

5.1 Encoding Boundaries by Digital Straight Segments

(About 56% of the Sects. 5.1 and 5.2 were quoted from the author's book [15] with the permission of the publisher.)

We have described in Sects. 4.2 and 4.3 algorithms for encoding the boundaries of homogeneous subsets. A homogeneous subset is a connected subset of the set of pixels in which all pixels contain the same value, a color, or an index. These algorithms are usable both for color images as well as for indexed images. However, it is important to throw the attention of the reader to the following property of color images: almost each homogeneous region in a color image consists of a single pixel. The reason is the extremely high number of different colors, namely 16,777,216 colors. Therefore, two pixels with an exceedingly small difference of one of their color channels, Red, Green, or Blue, are encoded as pixels with different colors. Thus, in the most cases each of the eight neighbors of a pixel has a color different from the color of the pixel. The pixel becomes a homogeneous region consisting of the single pixel.

If the detection and the encoding of the boundaries of homogeneous regions is to serve for the purpose of analyzing the structure of the image and especially for shape

Fig. 5.1 Example of a color image used in experiments

recognition, then it is not rational to use color images. One obtains much better results when the color image is first transformed into an indexed image since an indexed image contains homogeneous regions of essentially greater size than the corresponding color image.

Thus, for example about 75% of the loops of the encoded color image of 1600×1200 pixels presented in Fig. 5.1 are the smallest loops of 4 cracks containing a single pixel. All loops (except the single loop describing the frame of the whole image) are shorter than 30 cracks. Thus no one loop in this true color image contains more than 50 pixels.

We have encoded the loops of the boundaries with a list of the structures "Loops" and the array "Directions". The array "Directions" contains a number between 0 and 3 for each crack of a loop. This method of coding is simple, but it is not the most economical. It is possible to store four directions in one byte. Then the length of the field "Directions" would be about four times smaller. However, when using this way of encoding it is necessary to regard the possibility that a byte at the end of the loop can be filled not completely: It can contain less than four directions. To detect this situation, it is necessary to save for each loop the number of cracks (or directions) contained in it.

It would be desirable for the purpose of analyzing the structure of images and especially for shape recognition to break the loops into straight line segments and store the parameters of those segments. One of the possibilities to break a loop into straight line segments is the recognition of digital straight segments described in [15], Section 7.

5.1.1   Digital Straight Segments

Many authors [1, 2, 3, 4, 5, 6] have defined digital straight segments as sequences of pixels. A team around J.-P. Reveillès [5] has also developed algorithms for the segmentation of digital curves into digital straight segments. This is, however, not correct since pixels are elementary areas while digital curves are lines with area equal to zero. We suggest defining digital straight segments as subsets of lines thus consisting of cracks and points.

It is possible in digital geometry to use equations and inequalities to specify subsets of the space in the same way as in the classical analytical geometry. In the actual presentation, we write "coordinates" for combinatorial coordinates and "space" for an AC complex. Let us introduce some additional definitions, important for the future.

We define a digital straight segment (DSS) as a connected subset of the frontier of a half-space. The notion of "frontier" is widely used in topological publications. The difference between the notions of frontier and boundary, as already mentioned in Chap. 2, is the following: Frontier of a subset S of the space R is the set of all cells whose smallest neighborhood crosses the subset S and its complement $R - S$. Boundary of the subset S is the closure of the set of all 1-cells of R whose smallest neighborhood contains exactly one 2-cell of S. This difference is less important for our presentation.

A *closed digital half-space* of a two-dimensional space is the closure of the set of all 2-cells, whose coordinates satisfy a linear inequality. The complement of a closed half-space is an *open half-space*. A *digital half-plane* is a half-space of a two-dimensional space.

Figure 5.2 shows an example of the half-plane defined by the inequality $2x - 3y + 3 \geq 0$. All pixels of the half-plane are represented by shaded squares.

The cracks of the DSS composing the frontier of the half-plane are drawn in Fig. 5.2 as bold lines. The points of the DSS are drawn as small black disks.

As already mentioned above, we consider DSSs as sequences of cracks and points while in earlier publications [1, 2, 3, 4, 5, 6] they were considered as sequences of pixels. In practice, DSSs are mostly used in image analysis rather than in computer graphics. Therefore, considering them as sequences of cracks and points is more suitable for applications.

We need for the sequel the following definitions:

A cell c is said to be *strictly collinear* with two other cells a and b if the following condition holds:

Fig. 5.2 Examples of a
half-plane and of a DSS

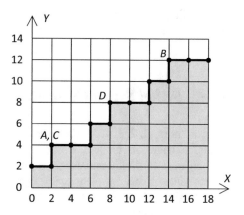

$$(x_b - x_a) \cdot (y_c - y_a) - (x_c - x_a) \cdot (y_b - y_a) = 0$$

The cell c is said to lie *to the positive side* of the ordered pair of cells a and b if

$$(x_b - x_a) \cdot (y_c - y_a) - (x_c - x_a) \cdot (y_b - y_a) > 0$$

It lies *to the negative side* of cells a and b if

$$(x_b - x_a) \cdot (y_c - y_a) - (x_c - x_a) \cdot (y_b - y_a) < 0$$

It seems perhaps that it would be better to speak about the right- and left-hand sides rather than about the positive and the negative ones. The reason to prefer the later notation is that the above definitions and inequalities are applicable without changes for both the mathematical coordinate system having the Y-axis directed upwards and to the coordinate system of computer graphics and image processing having the Y-axis directed downwards. The notions of right and left change when getting over from one system to the other; the notions of positive and negative do not. The above inequalities are constructed in such a way that the rotation of the segment (a, b) towards the segment (a, c) lying to the positive side of (a, b) corresponds to the positive direction of rotation usual in mathematics, i.e., from the positive direction of the X-axis to the positive direction of the Y-axis.

5.1.2   *Properties of Digital Straight Segments*

We consider here digital straight segments (DSSs) rather than digital lines because in practice one has always to do with finite subsets of a digital plane. Such a subset can only contain a subset of a line, i.e., a line segment. We shall investigate the conditions satisfied by the *combinatorial coordinates* of cells of a DSS. We prefer to do this for the combinatorial rather than for the standard coordinates because the derivations and the proofs are simpler. These conditions are different from those in standard coordinates since the parity of combinatorial coordinates of a cell depends on its dimension and on the orientation of a 1-dimensional cell. However, the deduced most important formulae are applicable both for combinatorial and for standard coordinates as explained below.

Standard coordinates are the coordinates commonly used in image processing and in computer graphics: the x-coordinate is the index of a column, and the y-coordinate is the index of the row in a rectangular grid, and this grid should be considered as a 2D Cartesian complex. Pixels are explicitly represented in the grid; however, the 0- and 1-cells are present implicitly.

Coordinate Assignment Rule   Each pixel F of a 2D image gets one 0-cell assigned to it as its "own cell". This is the 0-cell P lying in the corner of F which is the nearest to the origin of the coordinates. Two 1-cells which are incident to F and to P are also declared to be "own cells" of F. Thus, each pixel gets three own cells of lower dimensions. *All own cells of F get the same standard coordinates as F.* They can be distinguished by their type.

The transformations from combinatorial to standard coordinates are performed by the integer division by 2:

$$standard = combinatorial \ DIV \ 2;$$

where DIV denotes the integer division without rounding. This is a non-linear transformation. Therefore, not all properties of a DSS may be easily transferred from combinatorial to standard coordinates. We shall indicate this at the proper place.

To investigate the properties of DSSs it is easier and more comprehensible to consider at first the 2-cells (pixels) incident to the cells of a DSS rather than the cells of the DSS themselves. Consider a digital boundary curve K in a two-dimensional Cartesian space. It is possible to assign an orientation to K and thus to the 1-cells of K. Suppose that K does not intersect the border of the space. Then each 1-cell C of K is incident to exactly two pixels. One of them lies to the positive side (see the definition above) and the other to the negative side of the ordered pair of the endpoints of C.

The pixel P which is incident to an oriented crack C of the curve K and lies to the positive side of the ordered pair of the endpoints of C is called a *positive pixel* of K. Similarly, the incident pixel lying to the negative side of the ordered pair of the endpoints of C is called a *negative pixel* of K.

The set of all positive pixels of K will be called the *positive pixel set* of K and denoted by $SP(K)$. The set of all negative pixels of K will be called the *negative pixel set* of K and denoted by $SN(K)$.

If K is a DSS, then it is easily seen from the definitions of a DSS and of a closed digital half-plane that either $SP(K)$ belongs to a closed half-plane and $SN(K)$ belongs to its complement, or vice versa. This means that there is a linear form $H(x, y)$ such that $H(x, y) \geq 0$ for all positive pixels of K and $H(x, y) < 0$ for the negative ones.

There also exits another linear form $H'(x, y)$ (differing from $H(x, y)$ by a constant) such that $H'(x, y) > 0$ for all positive pixels of K and $H(x, y) \leq 0$ for the negative ones. We shall call both linear forms $H(x, y)$ and $H'(x, y)$ the *separating forms*.

The properties of a DSS, which we are interested in, and which are important for the recognition of a DSS, are known from the literature [3] during many years. However, we shall deduce these properties anew since most publications consider *sequences of pixels in standard coordinates* and we are interested in *sequences of cracks in combinatorial coordinates*. This is due to the importance of sequences of cracks as mentioned above. Besides that, the investigation by means of combinatorial coordinates is simpler than that by means of standard coordinates. The most important of our results are applicable for both standard and combinatorial coordinates due to using a parameter e denoting the minimum distance between two pixels. The value of e is equal to 1 in standard coordinates and to 2 in combinatorial ones.

We choose the signs of the coefficients of the separating form so that the form is non-negative for pixels lying to the positive side of the oriented cracks.

Now we define the important notions of the standard separating form and of the bases of a DSS:

The separating linear form $H(x. y) = a \cdot x + b \cdot y + r$ of a DSS D is called the *standard separating form* (SSF) of D if it satisfies the following conditions:

1. $H(x. y) \geq 0$ for all positive pixels of D and $H(x. y) < 0$ for all negative ones;
2. There are either at least two positive pixels P_1 and P_2 of D at which $H(x. y)$ takes its minimum value *with respect to* all positive pixels of D or at least two negative pixels N_1 and N_2 of D at which $H(x. y)$ takes its maximum value *with respect to* all negative pixels of D.
3. The coefficients of $H(x. y)$ are integers while a and b are mutually prime.

Definition BS: The set of positive pixels of a DSS D at which the SSF of D takes its minimum value *with respect to* all positive pixels of D is called the *positive base* of D. Similarly, the set of negative pixels of a DSS D at which the SSF of D takes its maximum value with respect to all negative pixels of D is called the *negative base* of D. Figure 5.3 shows an example of a DSS with its bases.

The pixel of the base B, which is the nearest to the starting point (respectively, endpoint) of an oriented DSS is called the *starting pixel* (respectively, the *ending pixel*) of B. Let S and E be coordinate vectors of the starting and the ending pixel of those base of the DSS, which contains more than one pixel. Then the vector parallel to E-S having mutually prime components is called the *base vector* of the DSS. If both bases contain a single pixel, which is only the case, when the DSS consists of a single crack, then the base vector is a unit vector parallel to that crack.

We shall formulate and prove in what follows a set of theorems concerning the properties of a DSS. These theorems are necessary to prove that the suggested algorithm for recognizing a DSS and for subdividing a given digital curve into as long as possible DSSs is correct.

First, we shall show that any DSS of a finite length possesses the standard separating form.

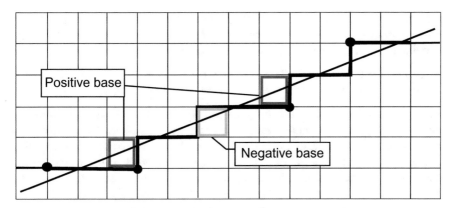

Fig. 5.3 A DSS (bold lines) and its bases

Theorem SSF For any DSS D in a Cartesian 2D-space there exists the *standard* separating form SSF. The absolute values of the coefficients a and b are not greater than the greatest dimension of the space. If D contains at least two cracks, then the SSF is *uniquely* defined.

Proof If D consists of a single crack C, $P = (x_P, y_P)$ and $N = (x_N, y_N)$ being the positive and the negative pixels incident to C, then $H(x. y) = a \cdot (x - x_P) + b \cdot (y - y_P)$ with $a = \text{sign}(x_p - x_n)$ and $b = \text{sign}(y_p - y_n)$ is obviously the SSF of D.

Let now D consist of at least two cracks. Since D is a DSS, there exists a separating linear form $h(x. y) = P \cdot x + Q \cdot y + R$ with real coefficients P, Q such that $h(x. y) \geq 0$ for $(x. y) \in SP$ and $h(x. y) < 0$ for $(x. y) \in SN$. Let $P_1 \in SP$ be the pixel with the property $h(P_1) = \min h(p)$ with respect to all pixels of SP. The linear form $h'(P_1) = h(p) - h(P_1) = a \cdot (x - x_1) + b \cdot (y - y_1)$ with x_1, y_1 being the coordinates of P_1, is a separating one and it has the value 0 at P_1. If there is one more pixel $P_2 \in SP$ with the property $h'(P_2) = 0$, then the conditions (1) and (2) of Definition SSF are fulfilled since $h'(P)$ may be written in the form $h'(P) = \pm ((y_2 - y_1)(x - x_1) - (x_2 - x_1)(y - y_1))/d$ where d is the greatest common divisor of $(y_2 - y_1)$ and $(x_2 - x_1)$. The coefficients $(x_2 - x_1)/d$ and $(y_2 - y_1)/d$ of this form are integers and their absolute values are obviously not greater than the greatest dimension of the space. Thus, all conditions of the theorem are fulfilled.

If P_1 is the only positive pixel with $h'(P) = 0$, then there are also no negative pixels with this property since $h(N) < h(N) < 0$ for $N \in SN$. Let us rotate the vector (a, b) of $h'(P)$ until the value of $h'(P)$ becomes 0 at some pixel P_2. The form $h'(P)$ remains separating during this rotation since $h'(P_1) = 0$, $h'(P) > 0$ for all positive and $h'(P) < 0$ for all negative pixels until P_2 is hit. If $P_2 \in SP$, then the conditions of the theorem are fulfilled for the same reason as in the case above. If, however, $P_2 \in SN$, then rotate the vector (a, b) of $h'(P)$ (while $h'(P_1) = 0$) in the opposite direction until the value of $h'(P)$ becomes 0 at some pixel P_3. If $P_3 \in SP$, then the conditions of the theorem are fulfilled. Otherwise rewrite $h'(P)$ as $h''(P) = a' \cdot (x - x_3) + b' \cdot (y - y_3)$ with the rotated values (a', b') of (a, b) and rotate the vector (a', b') in such direction that the value $h''(P_1)$ increases until $h''(P)$ becomes 0 at some pixel P_4. Since $h''(P)$ remains separating during the rotation, $P_4 \in SN$. The pixels P_3 and P_4 fulfill then the conditions of the theorem. All possible variants are exhausted.

Theorem TWD An oriented DSS D in a Cartesian complex contains oriented 1-cells of at most two directions which are not opposite to each other.

Proof Oriented cracks in a 2D Cartesian complex may have at most four different directions. Suppose D contains cracks of three different directions. Then one of them must be opposite to some other. Let it be the cracks C_1 and C_2. The corresponding ordered pairs of incident pixels also have opposite orientation. Let us denote the coordinates of a positive pixel incident to C_i, $i = 1,2$; by the vector PP_i and that of a negative one by the vector NP_i. Then $PP_1 = NP_1 + V$ and $PP_2 = NP_2 - V$. If the separating linear form $h(P)$ of D has at PP_1 a greater value than at NP_1, then $h(V) > 0$ and hence $h(PP_2) = h(NP_2) - h(V) < h(NP_2)$ which contradicts to the assumption that there is a separating linear form of D.

The direction of a crack that is opposite to one of the two directions of the cracks of a DSS is called a *prohibited* direction for this DSS.

Theorem SING In an oriented DSS D one of the directions of the cracks occurs singularly, i.e., each 1-cell of that direction is immediately followed by a 1-cell of the other direction. The singular direction is vertical if $|a| \leq |b|$ and it is horizontal if $|a| \geq |b|$. In the case $|a| = |b| = 1$ both directions are singular.

Proof Consider a digital curve of two horizontal and two vertical cracks as shown in Fig. 5.4. Its positive pixels are labeled by squares and the negative ones by circles. To separate the pixels incident to both horizontal cracks the linear form must have a slope $|a|/|b|$ *less than one*. However, to separate both pixels of the vertical cracks the same form must have a slope *greater than one*, which is impossible. Any digital curve having no singular direction must obviously contain a subset like that of Fig. 5.4, thus it cannot be a DSS.

The condition $|a| \leq |b|$ corresponds obviously to a DSS with a slope less or equal 1. In this case, the singular direction is a vertical one. In the case of $|a| \geq |b|$

Fig. 5.4 To the proof of Theorem SING; explanations in text

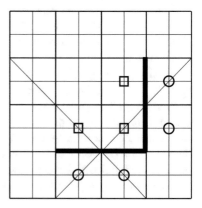

Fig. 5.5 To the proof of
Theorem INC; explanations
in text

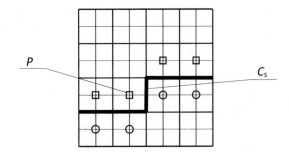

the slope is greater or equal to 1 and the pixels incident to two adjacent vertical
cracks may be separated. Thus, the singular direction is a horizontal one. In the case
$|a| = |b|$ the pixels incident neither to two adjacent horizontal nor to two adjacent
vertical cracks may be separated. Thus, both directions must be singular.

Let us denote the non-singular direction of the cracks of a DSS D as its *main*
direction. If both directions of D are singular, then any one of them may be chosen as
the main one.

Theorem INC Each positive pixel of a DSS D is incident to a crack of the main
direction. This crack is incident to a negative pixel of D.

Proof Consider a DSS D in a space with combinatorial coordinates. According to
Definition PP, each positive pixel of D is incident to some crack of D. We shall
demonstrate that if a positive pixel is incident to a crack of a singular direction, then
it is also incident to a crack of the other, orthogonal direction.
Let P be a positive pixel incident to a crack C_s of the singular direction. Let the
coordinates of C_s be (x, y) while one of the coordinates is odd and the other even
(as it is the case for any crack). Since the direction of C_s is singular, there must be in
D two cracks adjacent to C_s of an orthogonal direction. Let us denote them by C_1 and
C_2. Both coordinates x and y of these cracks differ from the coordinates of C_s by ± 1.
Since one of the coordinates of P differs by one from the even coordinate of C_s, this
coordinate of P is equal to the odd coordinate of either C_1 or C_2. Thus, P is incident
either to C_1 or to C_2. This crack is incident to a negative pixel of D as any crack of D.

Theorem VAL Let $H(x, y) = a \cdot x + b \cdot y + r$ be the standard separating form of the
DSS D and (x, y) are combinatorial or standard coordinates of a positive pixel of D.
Then the values of $H(x, y)$ for all positive pixels of D lie in the closed interval $[0, e \cdot
(\max(|a|, |b|)-1)]$, where $e = 2$ for combinatorial coordinates and $e = 1$ for standard
coordinates. The values of $H(x, y)$ for all negative pixels of D lie in the closed
interval $[-e \cdot \max(|a|, |b|), -e]$.

Proof The standard separating form H separates the pixels in such a way that $H(x,
y) \geq 0$ for all positive and $H(x, y) < 0$ for all negative pixels. Consider first the case
when the slope of D is positive and less than one, i.e., $a < 0; b > |a|$. Then the main

direction of D is the horizontal one. Let $P_1 = (x_1, y_1)$ be a positive pixel of D with the maximum value of $H(x, y)$ with respect to all positive pixels of D:

$$H(P_1) = H(x_1, y_1) = \max H(x, y) = M > 0$$

According to Theorem INC P_1 is incident to some crack of the main direction and the pixel $N_1 = (x_1, \ y_1 - e)$ is a negative one. Since $\partial H / \partial y = b$, $H(x_1, y_1 - e) = M - e \cdot b < 0$. Thus $\max H(x, y) = M < e \cdot |b|$.

In a similar way one may find the maximum value of $H(P_1)$ for a DSS with any other slope: e.g. if the slope is greater or equal 1, then $a < 0$; $0 < b \le |a|$ and the main direction is the vertical one. A positive pixel P_1 is incident to a vertical crack, $\partial H / \partial x = a$, $H(x_1 - e, \ y_1) = M - e \cdot |a| < 0$. Thus $\max H(x, y) = M < e \cdot |a|$ with respect to all positive pixels of D. Thus, for any slope

$$\max H(x, y) < e \cdot \max (|a|, |b|)$$

for all positive pixels of D. Since the differences $(x - x_0)$ and $(y - y_0)$ of the coordinates of any pixel (x, y) and a pixel (x_0, y_0) of the positive base (see Definition BS above) are multiples of e, so are also the values of $H(x, y)$.

Thus, $0 \le H(x, y) \le e \cdot (\max(|a|, |b|) - 1)$ for the positive pixels.

According to Theorem INC, for any positive pixel P of D there is a negative pixel N of D while both P and N are incident to a crack C of the main direction of D. The direction of C is horizontal if $|a| < |b|$ and the pixels P and N incident to C have different Y-coordinates. The values $H(P)$ and $H(N)$ obviously differ by $e \cdot |b|$ which value is equal to $e \cdot \max (|a|, |b|)$ in the case of $|a| < |b|$. If, however, $|a| > |b|$, then the main direction is vertical and the values $H(P)$ and $H(N)$ obviously differ by $e \cdot |a| = e \cdot \max (|a|, |b|)$. Thus, the values $H(P)$ and $H(N)$ differ in both cases by $e \cdot \max (|a|, |b|)$, which is also trivially true for the remaining case $|a| = |b|$.

This fact gives us the possibility to deduce the limits of $H(x, y)$ for the negative pixels from that of the positive ones simply by subtracting the value of $e \cdot \max (|a|, |b|)$.

Thus, $0 - e \cdot \max (|a|, |b|) \le H(N) \le e \cdot (\max(|a|, |b|) - 1) - e \cdot \max (|a|, |b|)$ or

$$-e \cdot \max (|a|, |b|) \le H(N) \le -e$$

for the negative pixels.

Theorem ALL Let $H(x, y) = a \cdot x + b \cdot y + r$ be the standard separating form of the DSS D and $(x, \ y)$ are combinatorial or standard coordinates of a positive pixel of D. Then $H(x, \ y)$ takes at positive pixels of D *all values* between 0 and $e \cdot (\max(|a|, |b|) - 1)$, which are multiples of e. It takes at negative pixels of D *all values* between $-e \cdot \max (|a|, |b|)$ and $-e$, which are multiples of e.

Proof It is easily seen that $H(x, y)$ is periodical with the two-dimensional period $e \cdot (b, -a)$. Indeed,

$H(x + e \cdot b, \ y - e \cdot a) = a \cdot (x + e \cdot b) + b \cdot (y - e \cdot a) + r = a \cdot x + e \cdot a \cdot b + b \cdot y - e \cdot b \cdot a + r = a \cdot x + b \cdot y + r = H(x, y)$
and if (x, y) is a pixel, then $(x + e \cdot b, \ y - e \cdot a)$ is again a pixel.

A period obviously contains $|b|$ horizontal and $|a|$ vertical cracks. According to Theorem SING there are exactly $\max(|a|, |b|)$ cracks of the main direction among them and hence exactly as many positive pixels. Suppose there are two positive pixels P_1 and P_2 inside of one period having equal values of $H : H(P_1) = H(P_2)$ or $a \cdot x_1 + b \cdot y_1 + r = a \cdot x_2 + b \cdot y_2 + r$. It follows $a \cdot (x_2 - x_1) + b \cdot (y_2 - y_1) = 0$ or

$$a \cdot (x_2 - x_1) = -b \cdot (y_2 - y_1) \tag{5.1}$$

Since $(y_2 - y_1)$ is integer, the left side of (5.1) is a multiple of b. However, a and b are mutually prime. Thus $(x_2 - x_1)$ must be equal to $k \cdot b, \ k \neq 0$, which contradicts the assumption that P_1 and P_2 are inside of one period. Thus, there are no two pixels inside of one period having equal values of $H(\ )$. There are exactly $\max(|a|, |b|)$ positive pixels inside of one period and $H(\ )$ takes values from the closed interval $[0, \ e \cdot (\max(|a|, |b|) - 1)]$ on these pixels. There are $\max(|a|, |b|)$ different values in the interval, which are multiples of e, and all values of $H(\ )$ that it takes $\max(|a|, |b|)$ pixels of a period are different. Thus $H(\ )$ takes *all values* of $[0, e(\max(|a|, |b|) - 1)]$ which are multiples of e.

Theorem BB There are no pixels in the space between the prolonged bases of a DSS.

Proof According to Definition BS the SSF $H(P)$ takes the value 0 at each pixel of the positive base and the value $-e$ at each pixel of the negative base. The prolonged bases are sets of pixels having all the same value of $H(P)$. If P is a pixel lying between the prolonged bases, then the value $H(P)$ must be some intermediate value between 0 and $-e$. However, the value of $H(P)$ at any pixel is a multiple of e. There are no multiples of e between $-e$ and 0.

Theorem SYM If two vectors V_1 and V_2 correspond to some pixels of a grid, then the vector $V = 2 \cdot V_1 - V_2$, which is symmetrical to V_2 with respect to V_1, also corresponds to some pixel of the (prolonged) grid.

Proof A vector corresponds to a pixel if and only if its both coordinates may be represented as $e \cdot i + 1$ where i is some integer. Let $V_1 = (e \cdot i + 1, e \cdot j + 1)$ and $V_2 = (e \cdot a + 1, e \cdot b + 1)$ Then $V = 2 \cdot V_1 - V_2 = (x, y)$ with $x = 2 \cdot (e \cdot i + 1) - (e \cdot a + 1) = e \cdot (2 \cdot i - a) + 2 - 1 = e \cdot k + 1$ whith $k = 2 \cdot i - a$. A similar calculation may be made for y. Thus, V corresponds to some pixel.

Theorem CD (Common Divider) Consider a linear form $H(P) = H(x, y) = a \cdot x + b \cdot y + r$ whose integer coefficients a and b are mutually prime and two vectors $P_1 = (x_1, y_1)$ and $P_2 = (x_2, y_2)$. If $H(P_2) - H(P_1) = M \neq 0$, then the differences $dx = x_2 - x_1$ and $dy = y_2 - y_1$ of the coordinates of P_1 and P_2 have no common divider greater than $|M|$.

Proof Suppose dx and dy have a common divider Div. Than the difference $M = H(P_2) - H(P_1)$ also has Div as its divider. Since $M \neq 0$ it is obvious that $|M| \geq Div$.

Corollary JB (Joining the Bases) If two pixels $P_1 = (x_1, y_1)$ and $P_2 = (x_2, y_2)$ belong to different bases of a DSS, than the values $b = (x_2 - x_1)/e$ and $a = -(y_2 - y_1)/e$ are mutually prime.

Proof According to Theorem ALL the values of the SSF $H(P)$ of a DSS S on the pixels P_1 and P_2 belonging to different bases of S differ by e. Hence, according to Theorem CD, the differences $dx = x_2 - x_1$ and $dy = y_2 - y_1$ of the coordinates of P_1 and P_2 have no common divider greater than e. It follows that the half differences have no common divider greater than 1.

5.1.3   Recognition of a DSS During the Tracing

Consider a sequence of cracks (it may be a single crack) and their end-points, which sequence is a DSS D with the standard separating form $H(x, y)$ represented as a function of a pixel P as

$$H(P) = a{\cdot}(P.x - Sp.x) + b{\cdot}(P.y - Sp.y)$$

where Sp is the starting pixel of the positive base (Sp, Ep). Let the negative base of D be (Sn, En).

Let C be a crack adjacent to the last crack of the sequence and let C have one of the non-prohibited directions. If both the positive and the negative pixels of C have values of $H(x, y)$ which lie in the allowed intervals specified by Theorem VAL then the sequence $D \cup \{C\}$ is obviously a DSS with the same standard separating form as that of D. The only necessary change of the parameters of D is that, if one of the pixels incident to C lies on one of the bases of D, then the corresponding base must be prolonged until that pixel. This case can be recognized by the value of $H(x, y)$, which is either 0 or -e. The following Theorem determines the conditions under which a DSS D being prolonged by a crack C is a DSS with parameters *different* from those of D.

Theorem PRO If the positive pixel P of C has the value $H(P) = -e$, i.e., it lies on the negative base of D, then the sequence $D' = D \cup \{C\}$ is still a DSS, which is different from D. The positive base of D' is now (Sp, P) where Sp is the starting pixel of the old positive base. Its negative base consists of the single pixel En that is the ending pixel of the old negative base. The coefficients a and b of the new standard separating form are $a = -(P.y - Sp.y)/e$ and $b = (P.x - Sp.x)/e$. If, however, $H(P) < -e$, then the sequence $D \cup \{C\}$ is no more a DSS.

Similarly, if the negative pixel N of C has the value $H(N) = 0$, i.e., it lies on the positive base of D, then the sequence $D'' = D \cup \{C\}$ is still a DSS which is different from D. The negative base of D'' is (Sn, N) and its positive base consists of the single pixel Sp. The coefficients a and b of the new standard separating form are $a = -(N.y - Sn.y)/e$ and $b = (N.x - Sn.x)/e$. If, however, $H(N) > 0$, then the sequence $D \cup \{C\}$ is no more a DSS.

Proof In the case $H(P) = -e$, the new linear form

$$H'(Q) = -(P.y - Sp.y) \cdot (Q.x - Sp.x)/e + (P.x - Sp.x) \cdot (Q.y - Sp.y)/e$$

obviously has the value 0 at $Q = P$ and at the starting point $Q = Sp$ of the positive base and is the SSF of the new DSS $D' = D \cup \{C\}$. Really, the new positive base is (Sp, P). The pixel P lies on the old negative base. All negative pixels of D are collinear with or lie on the negative side of the old base. Thus, they lie also on the negative side of the new base. The pixel P lies on the negative side of the old positive base. All positive pixels of D *after* Sp lie on the positive side of (Sp, P).

This is, however, not true for the positive pixels *before Sp*: those of them lying between the prolonged new and old bases can lie on the negative side of the new positive base. All such pixels (if they exist) would lie in the right triangle T whose one leg is (Sp, V) while V is the pixel symmetrical to Ep with respect to Sp and whose hypotenuse lies on the prolonged new base (Fig. 5.6). The image of this triangle symmetric with respect to Sp lies partially between the old bases and partially on the old base. According to Theorem BB, the only pixels in the image triangle are Sp and Ep. According to Theorem SYM, the only pixels in the triangle T are the symmetric pixels, namely Sp and V. The pixel V does not belong to the DSS D. Otherwise it would be the starting pixel of the old positive base. Thus, there are no pixels of D lying on the negative side of the new base, and we can define the parameters of the linear form $H'(\ )$ as $a = -(P.y - Sp.y)/e$ and $b = (P.x - Sp.x)/e$.

The new positive base is (Sp, P) and the new negative base consists of a single pixel: the starting pixel Sn is set equal to the old ending pixel En. Since the pixels Sp and P lie on different bases of the old DSS D, the values of a and b as defined above are according to Corollary JB mutually prime. Thus the new linear form $H'(\ )$ is the standard one.

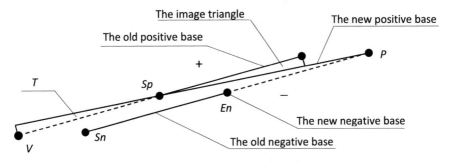

Fig. 5.6 To the proof of Theorem PRO; explanations in the text above

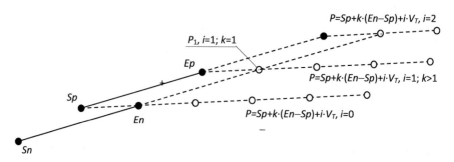

Fig. 5.7 To the proof of Theorem PRO

Consider now the case $H(P) < -e$. Since the coordinates of any two pixels differ by a multiple of e and $H(Sp) = 0$ the values of H on pixels are all multiples of e. Thus $H(P) < -e$ means $H(P) = -e \cdot k$ with $0 < k = 2, 3, \ldots$ (see Fig. 5.7).

We shall investigate the case when the positive base contains exactly two pixels: the starting pixel Sp and the ending pixel Ep. In this case, the drawing is clearer than in the case of a long positive base since the lines of the drawing become the closer to each other the longer the base. However, the consideration remains the same independently of the length of the base.

Let P be a positive pixel of C with $H(P) = -e \cdot k$, $k > 1$. Then P may be represented as $P = Sp + k \cdot (En - Ep) + i \cdot V_T$, where $V_T = (b, -a) = (Ep - Sp)$ is the base vector of D. The pixels with the coordinates $P = Sp + k \cdot (En - Sp)$ ($i = 0$) lie all on the prolonged line (Sp, En) (Fig. 5.7) and are collinear with Sp and En. Thus no such pixel as well as no pixel lying to the negative side of (Sp, En) can be used as the ending point of the new positive base: the pixel En would be not separated from Sp. However, pixels corresponding to $i > 0$ and $k > 1$ lie farther from Sp than the pixel P_1 with the same value of i and with $k = 1$ which pixel lies on the prolonged negative base. Thus, P_1 is a negative pixel of D, but it cannot be separated from Sp since it lies nearer to Ep than P and therefore it lies to the positive side of the pair (Sp, P). But in the case $H(P) < -e$ there exists no pixel which could serve as the ending point of the new positive base and hence $D \cup \{C\}$ is not a DSS.

The same considerations may be repeated for the case of an arbitrarily long base and for the cases of a negative pixel N with $H(N) = 0$ and $H(N) < 0$.

We suggest the following algorithm for subdividing a given simple digital curve CV into as long as possible DSSs and proof its correctness.

5.1.4   Algorithms for Subdividing a Line into Digital Straight Segments

We have called a digital straight segment (DSS) a connected subset of the boundary of a digital half-plane while a closed digital half-plane is the closure of the set of all pixels whose coordinates satisfy a linear inequality.

Now we describe the algorithms "ReadLoops", "TraceLoop" and "Reco". The algorithms are suitable for the graphical coordinate system, where the positive Y-axis is pointing downwards. If the reader uses the mathematical coordinate system, where the positive Y-axis is pointing upwards, it should replace 'left' by 'right' and 'right' by 'left' everywhere.

The algorithm "ReadLoops" is designed for processing the loops calculated by the algorithm "CORB" described in Sect. 4.2. However, a similar algorithm can be used for any closed lines consisting of cracks and points.

The algorithm "ReadLoops" reads all loops and calls the subroutine "TraceLoop" for a loop if the number of cracks in the loop is greater or equal to the predefined value "minLength". The algorithm "TraceLoop" calls the utility algorithm "Ini" which prepares the necessary data for the recognition of a DSS. Then "TraceLoops" reads one loop, reads the directions from the array "Directions", calculates coordinates of the next crack, and calls the algorithm "Reco" which checks whether the sequence of the cracks between the starting point and the actual crack is a DSS. It returns the logical value "true" if it is, or "false" otherwise.

To recognize the longest DSS starting at a given point, it is necessary to find a digital half-plane containing the maximum number of pixels lying immediately at the right side of the loop, called right pixels, and containing no pixel at the left side. Such a half-plane is called separating half-plane. Its boundary separates the right pixels from the left ones. It is described with the separating linear form H which has positive values or the value zero at all right pixels and negative values at left pixels.

Now we describe the algorithm "Reco". Its parameters are the two-dimensional vector "Crack" and its direction "dir". The recognition starts at the starting point of "Crack". Combinatorial coordinates of the right pixel R and left pixel L of "Crack" are calculated.

The sequence of cracks composing a DSS cannot contain a crack C_1 with direction opposite to the direction *dir* of "Crack" because there exist no half-plane separating the right and left pixels of "Crack" and those of C_1. If the algorithm "Reco" is at the first crack of the loop, then the first prohibited direction *Proh1* is set as opposite to *dir*. The second prohibited direction *Proh2* will be defined later. "Reco" calculates the parameters of the separating half-plane for the DSS consisting of a single crack and returns "false" as a sign that no end point of a DSS is reached,

At any other crack, the algorithm "Reco" checks whether "dir" is equal to one of the prohibited directions. If this is the case, then it returns "true" as the sign that the DSS is finished: The actual crack does not belong to the running DSS. If *dir* is not opposite to *Proh1* and *Proh2* is still unknown, then set *Proh2* as opposite to *dir*.

If "Reco" is at the second crack and *Proh2* is still unknown, then it returns "false": The first two cracks have the same direction, the parameters of the separating half-plane remain unchanged. If, however, *Proh2* is already known, then "Reco" calculates new values of the coefficients of the separating linear form and returns "false".

Subdivision of the loops into DSSs makes it possible to find longer straight-line segments in the image and estimate their location and direction which is useful for

Fig. 5.8 Binary image of a car used for calculating the DSSs

Fig. 5.9 End points of the DSSs longer than 200 cracks for the image of Fig. 5.8

the analysis of images. Thus, for example, Fig. 5.8 shows a binary image of a car and Fig. 5.9 shows the DSSs of loops longer than 200 cracks.

It is possible to recognize the wheels, some longer horizontal segments between the wheels, below the windows and at the top of the roof, which enables the recognition of a car.

Let us now present the algorithms for dissolving loops into DSSs. The following algorithms use the global array "Loop[]" of structures "CLoop". Each structure contains the variables "First", "Last", and the two-dimensional vector "Start" with the coordinates of the starting point.

Algorithm ReadLoops(minLength)

The variable iVert is global.

Step 1: Start.
Step 2: Declare variables first, iLoop, Length.
Step 3: Initialize global variable iVert ← 0.
Step 4: Initialize variable iLoop ← 0.
Step 5: Repeat until iLoop is less than nLoop:
    5.1: If iLoop is equal 0, then set first ← 0
        else set first ← Loop[iLoop - 1].Last + 1.
    5.2: Set Length ← Loop[iLoop].Last - first + 1.
    5.3: If Length is geater than minLength, then start TraceLoop(iLoop, first).
    5.4: Set iLoop ← iLoop + 1.   (End of the repetition Step 5)
Stop.

Below we describe the algorithm "TraceLoop(iLoop, first)".

The array of two-dimensional vectors "Vertex", the array "Directions" and the variable "iVert" are global. The algorithm "TraceLoop" traces the loop with the index "iLoop". It calculates the coordinates of subsequent cracks and calls for each crack the subroutine "Reco(Crack, dir)" described below. The recognition of a DSS starts at the first crack of the loop. The subroutine "Ini" calculates the starting values of the parameters of a DSS. It sets the global variable "iCrack" equal to 0 and the global variables "Prohibit1" and "Prohibit2" equal to −1. The subroutine "Reco" calculates for each actual crack the parameters of the DSS containing the sequence of cracks from the first crack of the loop until the actual crack. If this sequence is a DSS, then "Reco" returns the value "false" and the processing of the DSS can be continued. If, however, the sequence is not a correct DSS, then this means that the actual crack should not be accepted as an allowed continuation of the DSS. The DSS ends before the actual crack, at its starting point. The subroutine "Reco" returns the value "true". The end point of the DSS being the starting point of the actual crack is saved in the array "Vertex" which is a member of the class to which the subroutines "ReadLoops", "TraceLoop" and "Reco" belong. The start of the next DSS is prepared by calling "Ini" and "Reco" with the actual crack as its parameter.

Algorithms "TraceLoop" and "Reco" use the small array "Step[ ]" of vectors defined as follows:

$Step[0] = (1, 0)$; $Step[1] = (0, 1)$; $Step[2] = (−1, 0)$; $Step[3] = (0, −1)$;

Algorithm TraceLoop (iLoop, first)

The variables StartPoint, iVert, the arrays Directions, Loop, Vertex, and
the array Step of four two-dimensional vectors are global.

Step 1: Set iVert. ← 0.
Step 2: Declare integer variables dir, first, iDir, and the logical variable break.
Step 3: Initialize the variable Vertex[iVert] ← Loop[iLoop].Start.
Step 4: Set iVert ← iVert + 1.
Step 5: Declare two-dimensional vectors Crack and Point.
Step 6: Set Point ← 2* Loop[iLoop].Start.
Step 7: Call Ini().
Step 8: If iLoop is equal to 0, then set first ← 0.
        else set first ← Loop[iLoop - 1].Last + 1.
Step 9: Initialize the variable iDir ← first.
Step 10: Repeat until iDir is less or equal to Loop[iLoop].Last:
        10.1: Set dir ← Directions[iDir].
        10.2: Set Crack ← Point + Step[iDir].
        10.3: Call break ← Reco(Crack, dir).   (Reco returns true or false)
        10.4: If break is true, then
              Begin
                Set StartPoint ← Point.
                Set Vertex[iVert].X ← Point.X / 2. (End point of the DSS)
                Set Vertex[iVert].Y ← Point.Y / 2. (Standard coordinates)
                iVert ← iVert + 1.
                call Ini().
                call Reco(Crack, dir).
              End
        10.5: Set Point ← Crack + Step[dir].
        10.6: Set iDir ← iDir + 1.   (End of repetition step 10.)
Step 11: Set Loop[iLoop].LastVert ← iVert - 1.
Stop.

We describe now the algorithm "Reco(Crack, dir)":

The algorithm "Reco" uses the following global arrays of two-dimensional
vectors:

Step[4] which has been defined above and

ToRight[0] = (0, 1); ToRight [1] = (−1, 0); ToRight [2] = (0, −1); ToRight
[3] = (1, 0);

It also uses the method "Opposite(dir)" returning a direction opposite to the
direction "dir" and the method "GetH(vector V)" returning the value of the linear
form a*(V.X − StartR.X) + b*(V.Y − StartR.Y). The algorithm "Reco" also uses
global two-dimensional vectors EndL, EndR, StartL, StartR and global variables a,
b, e, iCrack, Prohibit1, and Prohibit2.

Algorithm Reco(Crack, dir):

The global variables e, iCrack, Prohibit1, and Prohibit2 are set by Ini(). Refer to explanations in the above text.

Step 1: Declare variables L, R, EndL, EndR, StartR, and StartL as two-dimensional vectors.
Step 2: Declare integer variables HR and HL.
Step 3: Initialize variable R ← Crack + ToRight[dir].
Step 4: Initialize variable L ← Crack - ToRight[dir].
Step 5: If iCrack is equal to 0, then
    Begin
    Set Prohibit1 ← Opposite(dir).
    Set Prohibit2 ← -1.
    Set iCrack ← 1.
    Set StartR ← EndR ← R.
    Set StartL ← EndL ← L.
    Set a ← -Step[dir].Y.
    Set b ← Step[dir].X.
    return false.
    End
Step 6: If dir equals to Prohibit1 or dir equals to Prohibit2, then return true.
Step 7: If dir is not equal to Prohibit1 and Prohibit2 equals -1, then
                            set Prohibit2 ← Oposite(dir).
Step 8: If iCrack is equal to 1, then
      Begin 8
      Set EndR ← R.
      Set EndL ← L.
      Set iCrack ← 2.
      If Prohibit2 is equal to -1, then return false.
      If EndR is equal to StartR, then
       Begin
        Set a ← -(EndL.Y - StartL.Y) / e.
        Set b ← (EndL.X - StartL.X) / e.
       End
       else
       Begin
        Set a ← -(EndR.Y - StartR.Y) / e.
        Set b ← (EndR.X - StartR.X) / e.
       End
       return false.
      End 8   (End of If iCrack is equal to 1, step 8.)
Step 9: Initialize HR ← GetH(R).
Step 10: Initialize HL ← GetH(L).
Step 11: If HR < -e or HL > 0, then return true.
Step 12: If HR equals 0, then set EndR ← R.
      else
     if HR equals -e, then
      Begin
       Set EndR ← R.

Set StartL ← EndL.
            Set a ← -(EndR.Y - StartR.Y) / e.
            Set b ← (EndR.X - StartR.X) / e.
        End
Step 13: If HL equals -e, then set EndL ← L.
        else
            if HL equals 0, then
                Begin
                    Set EndL ← L.
                    Set StartR ← EndR.
                    Set a ← -(EndL.Y - StartL.Y) / e.
                    Set b ← (EndL.X - StartL.X) / e.
                End
Step 14: return false.
Stop

Figure 5.10 shows the indexed image of a wheel of a car, an example of the detected digital straight segments, and of a circle calculated by means of a modified method of least squares (Sect. 5.4.4).

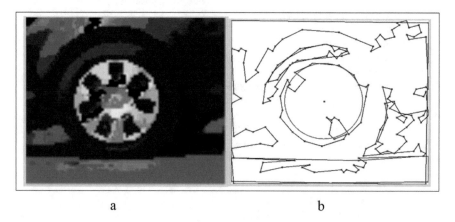

a   b

Fig. 5.10 The indexed image of a wheel of a car (**a**) and the DSSs representing the loops with the circle calculated for one of the loops (**b**)

5.1.5 Inequalities for Cracks and Points of a DSS in Combinatorial Coordinates

The values of the SSF at cells of lower dimensions underlie to regularities, which are different from those of the values at pixels. They are less important for the recognition of DSSs. However, they are also of theoretical interest, because, as it will be seen, they are the theoretical foundation of the well-known notions of "naive" and "standard" DSSs [7].

Theorem VC (Value at Crack) Let $H(z) = H(x, y) = a \cdot x + b \cdot y + r$ be the SSF of the DSS D, x and y being the combinatorial coordinates of the cell z. Let C be a crack of D having the main direction. Then $H(C) = H(N) + max\ (|a|, |b|) = H(P) - max\ (|a|, |b|)$, where N is the negative and P the positive pixels incident to C. If C has the singular direction, then $H(C)=H(N) + min\ (|a|, |b|) = H(P) - min\ (|a|, |b|)$. For a horizontal crack C_h, $H(C_h) = H(N) + |b| = H(P) - |b|$. For a vertical crack C_v $H(C_v) = H(N) + |a| = H(P) - |a|$.

Proof Let the crack C be a horizontal one and P be a pixel incident to C. Then the combinatorial coordinate x of C and P are equal, and the combinatorial coordinate y of C differs from that of P by 1. Therefore, the values $H(C)$ and $H(P)$ differ by $|b|$. The same is true for the values $H(C)$ and $H(N)$ where N is another pixel incident to C. If P is a positive and N the negative pixel of D, then according to the definition SSF $H(P) > H(N)$. Since $H(\ )$ is linear, the value $H(C)$ lies between $H(P)$ and $H(N)$: $H(C) = H(N) + |b| = H(P) - |b|$. According to Theorem SING the non-singular cracks of a DSS are horizontal if $|b| \geq |a|$. Thus, in the case when the main direction is horizontal $H(C) = H(N) + \ max\ (|a|, |b|) = H(P) - \ max\ (|a|, |b|)$. If the crack C is vertical, then the values $H(C)$ and $H(P)$ differ by $|a|$: $H(C) = H(N) + |a| = H(P) - |a|$. In this case $|a|$ is greater or equal to $|b|$. Thus, also in this case $H(C) = H(N) + \ max\ (|a|, |b|) = H(P) - \ max\ (|a|, |b|)$. If C has the singular direction, then it is orthogonal to the main direction and the differences $H(C)-H(N)$ and $H(P)-H(C)$ are equal to the absolute value of the coefficient of $H(\ )$ other as $max(|a|, |b|)$ which is $min(|a|, |b|)$.

Each crack C of a DSS is incident to a positive P and to a negative N pixel. According to Theorem VC the value $H(C_h)$ of the SSF for a horizontal crack C_h satisfies the equation:

$$H(C_h) = H(P) - |b| = H(N) + |b|;$$

The minimum value of $H(C_h)$ is achieved by the horizontal crack incident to the positive pixel having the minimum value of $H(P)$, which is equal to 0. Therefore, the minimum value of $H(C_h)$ is $-|b|$. Its maximum value is achieved by the horizontal crack incident to the negative pixel having the maximum value of $H(N)$ which is equal to $-e$. Therefore, the maximum value of $H(C_h)$ is $|b|-e$. Thus, the value of $H(C_h)$ satisfies the inequalities:

$$-|b| \leq H(C_h) \leq |b| - e.$$

It may be demonstrated in a similar way that the value of $H(C_v)$ for a vertical crack satisfies the inequalities:

$$-|a| \leq H(C_v) \leq |a| - e.$$

The value of $H(C_s)$ for a crack C_s of the singular direction satisfies the inequalities:

$$\min (|a|, |b|) \leq H(C_s) \leq \min (|a|, |b|) - e;$$

and for a crack C_n of the non-singular direction the inequalities:

$$\max (|a|, |b|) \leq H(C_n) \leq \max (|a|, |b|) - e.$$

These inequalities are guilty if a crack is defined by its combinatorial coordinates. The inequalities for standard coordinates depend upon the definition of the standard coordinates of cells of lower dimensions and will not be presented here.

The derivation of the inequalities for the points of a DSS is more complicated. A point of a DSS, as any other point in a two-dimensional Cartesian space, is incident to four pixels. There are among them two diagonal pairs of pixels. "Diagonal" means that *both* coordinates x and y of the two pixels of a diagonal pair differ by the value of e. One of these pairs is called the *main diagonal pair*. This pair must contain a positive P and a negative N pixel, while the difference $H(P) - H(N)$ is greater or equal to the similar difference of the other diagonal pair.

Definition MD The vector MD with the following properties is called *the main diagonal of a DSS*: its x-component is equal to 1 if the coefficient a of the standard separating form is *greater* or equal 0 and it is equal to -1 otherwise. Similarly, its y-component is equal to 1 if the coefficient b of the standard separating form is *greater* or equal 0 and is equal to -1 otherwise. If both a and b are equal to 0, then the vector MD is undefined.

It is easily seen that the main diagonal always composes an acute angle with the gradient of the standard separating form.

Theorem PV (Point Values) The values of the SSF $H(Po)$ of a DSS at its points Po satisfy the following inequalities:

$$-(|a| + |b|) \leq H(Po) \leq |a| + |b| - e.$$

Proof Consider the positive pixel P_{min} of the DSS having the minimum value of the SSF. According to Definition SSF $H(P_{min}) = 0$. The main diagonal MD composes an angle with the gradient of $H(\cdot)$, which is smaller than the angle of the other diagonal. Therefore, the point with the coordinates $(x' - MD_x, y' - MD_y)$ has the smallest value of $H(Po)$ among the four points incident to P_{min}. Thus, this is the minimum value of $H(Po)$ with respect to all points of the DSS. This value is equal to:

$$\min H(Po) = a \cdot (x' - MD_x - x_0) + b \cdot (y' - MD_y - y_0)$$
$$= a \cdot (x' - x_0) + b \cdot (y' - y_0) - a \cdot MD_x - b \cdot MD_y.$$

According to Definition MD, $a \cdot MD_x = |a|$ and $b \cdot MD_y = |b|$, while $a \cdot (x' - x_0) + b \cdot (y' - y_0) = H(P_{min}) = 0$. Consequently, $\min H(Po) = -(|a| + |b|)$.

Consider now the negative pixel $N_{max} = (x'', y'')$ having the maximum value of SSF. According to Theorem VAL $H(N_{max}) = -e$. Because of the above-mentioned properties of MD the point with the coordinates $(x'' + MD_x, y'' + MD_y)$ has the greatest value of $H(Po)$ among the four points incident to N_{max}. Therefore, this value is the maximum value of $H(Po)$ with respect to all points of the DSS. This value is equal to:

$$\max H(Po) = a \cdot (x'' + MD_x - x_0) + b \cdot (y'' + MD_y - y_0)$$
$$= a \cdot (x'' - x_0) + b \cdot (y'' - y_0) + a \cdot MD_x + b \cdot MD_y.$$

As stated above, $a \cdot MD_x = |a|$ and $b \cdot MD_y = |n|$, while $a \cdot (x'' - x_0) + b \cdot (y'' - y_0) = H(N_{max}) = -e$. Consequently, $\max H(N_{max}) = |a| + |b| - e$.

These inequalities are only guilty in combinatorial coordinates.

5.2   Method of Additional Integer Parameters of DSS

Following methods of encoding digital curves represented as sequences of DSSs are known:

The Crack Code
A well-known way of encoding digital curves is that of the Freeman code [8]. Boundary curves in a two-dimensional complex are sequences of cracks and points. Oriented cracks may have only four directions. Therefore, they may be encoded by a Freeman code with *four directions* that is also well known as the crack code (see e.g. [6]).

This way of encoding curves is a rather economical one especially if only two bits per crack are used. Its main drawback is the difficulty of performing geometrical transformations: only a translation is easy to realize. Rotation and scaling are hardly realizable without converting the crack code into some other code more suitable for geometrical transformations.

End Points of the DSSs
Another way of encoding curves consists in decomposing the curve into as long as possible DSSs and recording the coordinates of their endpoints. This code makes geometrical transformations easily realizable: it suffices to multiply the vectors

corresponding to all endpoints with the matrix of the desired transformation, eventually in homogeneous coordinates. However, this code does not allow an *exact* reconstruction of the original digital curve since there exist *many different DSSs* having the same endpoints. The distance between any two such DSSs is never greater than the pixel's diagonal. Therefore, the difference between such two DSSs may be considered in many applications as negligible. In such cases this way of encoding curves is the simplest and the most economical one, especially if we record coordinate increments rather than the coordinates themselves.

If, however, an exact reconstruction is necessary than the following ways of encoding are possible.

Floating Point Coordinates

A DSS is uniquely defined by its endpoints together with any Euclidean straight line that is its preimage. The Euclidean line parallel to the bases and lying exactly in the middle between them is such a line. We shall call the line *the axis of the DSS*. The parameters of the axis and the location of the endpoints may be combined if we calculate the crossing points of the axis with the main diagonals containing the endpoints. The crossing points uniquely define both the axis and the integer endpoints that are the integer points nearest to the crossing points. Note that the crossing point of an axis with the main diagonal of a DSS does not coincide in the general case with the crossing point of the axes of two adjacent DSSs. Therefore, it is necessary to save *two crossing points* for each endpoint of a DSS.

This way of encoding has the following drawbacks:

1. The memory demand is relatively great, e.g., one needs two "float" coordinates of 4 byte each for each of two crossing points which makes $4 \times 2 \times 2 = 16$ bytes per end-point;
2. A common endpoint of two subsequent DSSs of a digital polygon may be "split" after geometrical transformations into two different integer points thus making the polygon disconnected.

There are some rather complicated ways of overcoming one of the above drawbacks but not both.

We have suggested [9] a method overcoming the drawbacks.

Method of Additional Integer Parameters

A DSS may be uniquely specified by its endpoints and by one of its bases (Sect. 5.1.2). Specifying the base by the coordinates of its endpoints (which are mostly different from the endpoints of the DSS) is rather redundant. A more economical way of encoding consists in specifying two integers a and b specifying the first two coefficients of the SSF accompanied by the value L of the SSF at the starting point of the DSS. The parameters a, b and L together with the coordinates (x_s, y_s) of the starting point of the DSS uniquely specify the SSF:

$$H(x, y) = a \cdot (x - x_s) + b \cdot (y - y_s) + L.$$

The coordinates of the endpoints of the DSS may also be encoded economically: only the coordinates of the starting point of the whole curve must be specified explicitly; all other coordinates may be defined by means of the number NC of cracks in the current DSS. Thus, we need four integers per DSS: NC, L, a, and b. Since the majority of the DSSs are relatively short, these integers are mostly small ones: all four integers may be mostly packed into a word of 2 bytes. For longer DSSs a longer word of 3 to 8 bytes may be needed, but the frequency of such long words is low. Therefore, the average number of bytes per one DSS lies between 2 and 3.

This way of economically encoding the DSSs was successfully used for encoding quantized images without loss of information as described below. The details of the method and some experimental results are presented in [9].

In Sect. 5.1.4 we have described the algorithm "ReadLoops", which reads the codes of boundaries of regions, and subdivides them into longest possible DSSs. The output of the algorithm is a list of coordinates of the end points of the DSSs. That was the simplest version of the algorithm. Here we describe an advanced version of the algorithm that recognizes the longest DSSs during the tracing and simultaneously calculates the additional parameters described above which enable an *exact* reconstruction of the encoded image. The output of the algorithm is a list of codes, each of which contains economically encoded values of four integer parameters NC, L, a, and b. The parameter NC is the number of cracks in the DSS, L is the value of the standard separating form SSF (see Sect. 5.1.2) at the starting point of the DSS, a and b are the coefficients of the SSF specifying the slope and the orientation of the DSS. The coordinates of the starting point must be saved only for the first DSS of a line. They are saved in the structure of the starting branch point of the line in the cell list. The coordinates of the starting points of all other DSSs can be calculated by means of the four parameters during the reconstruction of the image.

The following section contains the algorithms of the subroutines "Reco_LMN" and "Encode_LMN". We present here no description and bring instead some comments to the most important lines of the algorithm in round brackets. It is, however, necessary to explain that due to the limited length of the code of a DSS only DSSs with the length less than or equal to 511 cracks and with the absolute values of the coefficients m and n less than or equal to 63 can be encoded. Therefore, the algorithm interrupts a DSS whose parameters have reached during the tracing of a boundary one of these critical values. Thus, a long DSS, e.g., at the boundary of the image, is subdivided into many shorter DSSs. This does not prevent the exact reconstruction of the image. The total length of the code increases only insignificantly.

5.2.1   Algorithm for Calculating the Additional Parameters

Variables StartPoint, StartPB, nCracks, Prohibit1, Prohibit2, EndPB, EndNB and the field ToPos, which are not declared in the algorithm of "Reco_LMN" are global.

Algorithm Reco_LMN(vector Crack, int dir)

The algorithm "Reco_LMN" uses the field Param[4] of vectors defined as Param[0] = (1, 0); Param[1] = (0, 1); Param[2] = (-1, 0); Param[3] = (0, -1).

Step 1: Declare and initialize variables MaxNM ← 63, MaxStep ← 511; (longer DSSs are subdivided)

Step 2: Declare and set the two-dimensional vector Dist ← Crack - StartPoint

Step 3: If |Dist.X|+|Dist.Y| is greater than 2*MaxStep, then return 1. (Too long)

Step 4: Set Pos ← Crack+ToPos[dir] and Neg ← Crack - ToPos[dir]. (Incident pixels)

Step 5: If nCracks is equal to 0, then   (nCracks is initialized by "ini()".)
 Begin
  Set Prohibit1 ← Opposite(dir).
  Set Prohibit2 ← -1.
  Set nCracks ← 1.
  Set StartPB ← EndPB ← Pos.
  Set StartNB ← EndNB ← Neg.   (Bases of the DSS)
  Set m ← Param[dir].Y.
  Set n ← Param[dir].X.   (These are the parameters of the DSS
  Set L ← (m*(StartPoint.X-EndNB.X+sign(m)) -NB.Y-sign(n)))/2.
  return 0.
 End

Step 6: If dir equals Prohibit1 or dir equals Prohibit2, then return 1. (Prohibited direction)

Step 7: If dir is not equal to Opposite(Prohibit1) and Prohibit2 equals-1, then
                                        Set Prohibit2 ← Opposite(dir).

Step 8: If nCracks equals 1, then
 Begin
  Set EndPB ← Pos.
  Set EndNB ← Neg. nCracks ← 2.   (Correcting the bases)
  If Prohibit2 equals -1, then return 0.   (Only one direction in the current DSS)
  If EndPB is equal to StartPB, then
  Begin
   Set m ← (EndNB.Y-StartNB.Y)/e.
   Set n ← (EndNB.X-StartNB.X)/e.
  End
  else
  Begin
   Set m ← (EndPB.Y-StartPB.Y)/e.
   Set n ← (EndPB.X-StartPB.X)/e.}
  Set L ← (m*(StartPoint.X-EndNB.X+sign(m)) -n*(StartPoint.Y-EndNB.Y-sign(n)))/2.
  return 0.   (Any two allowed cracks compose a DSS)
  End
 End   (end of If nCracks equals 1, step 8.)

Step 9: Define and initialize HP ← GetH(Pos) and HN ← GetH(Neg).
                                (Calculating two values of the SSF)

Step 10: If HP is less than -e or HN is greater than 0, then return 1. (Not a DSS)

Step 11: If HP equals 0, then set EndPB ← Pos.
        else
            If HP equals -e, then
            Begin
                Set EndPB ← Pos.
                Set StartNB ← EndNB.   (Correcting the bases)

If |(EndPB.Y-StartPB.Y)/e| is greater than MaxNM or
        |(EndPB.X-StartPB.X)/e| is greater than MaxNM, then return 1.
                                                  (Too long DSS)
      Set m ← (EndPB.Y-StartPB.Y)/e. n ← (EndPB.X-StartPB.X)/e.
   End
Step 12: If HN equals -e, then set EndNB ← Neg.   (Correcting the negative base)
    else
     If HN equals 0, then
     Begin
      Set EndNB ← Neg.
      Set StartPB ← EndPB.   (Correcting the bases)
      If |(EndNB.Y-StartNB.Y)/e| is greater than MaxNM or
       |(EndNB.X-StartNB.X)/e| is greater than MaxNM, then return 1. (Too long)
      Set m ← (EndNB.Y-StartNB.Y)/e.
      Set n ← (EndNB.X-StartNB.X)/e.
     End
Step 13: L ← (m*(StartPoint.X-EndNB.X+sign(m)) –
                        n*(StartPoint.Y-EndNB.Y-sign(n)))/2.
Step 14: return 0.
Stop

5.2.2   Algorithm for Encoding the Additional Parameters

The algorithm packs the parameter NC and the global variables L, m, n into 2, 3 or 4 Bytes, saves them in the variable "Code" and returns "Code". The parameter "Anz_Byte" must be defined in such a way, that it can be changed by the algorithm.

Algorithm Encode_LMN(NC, Anz_Byte)

The variables "L", "m", "n", and "NC" are global. They are calculated by the algorithm "Reco_LMN".

Step 1: Declare variables Code, absM, absN and typ.
Step 2: Declare and initialize fields: Max_Step[3] ← {31,127,511},
    Max_L[3] ← {7,31,127}, Max_NM[3] ← {3,15,63}, Anz_Byt[3] ← {2,3,4}.
    Shift_L[3] ← {9,11,13}, Shift_M[3] ← {12,16,20}, Shift_N[3] ← {14,20,26}.
Step 3: Set two variables: absM ← |m| and absN ← |n|.
Step 4: Set typ ← 0.
Step 5: Repeat until typ remains less than 3:   (defining the type of the code)
  Begin
    If NC is less or equal to Max_Step[typ] and L is less or equal to Max_L[typ]
                and absM is less or equal to Max_NM[typ]
                and absN is less or equal to Max_NM[typ], then break repetition.
  Set typ ← typ + 1.
  End
Step 6: Set Anz_Byte ← Anz_Byt[typ]. (Next steps pack the parameters into the code):
Step 7: Set Code ← typ.   ("type" stays in Bits 0 and 1 of Code)
Step 8: If m is negative, then set Code ← Code+4. (The sign of "m" stays in Bit 2)
Step 9: If n is negative, then set Code ← Code+8. (The sign of "n" stays in Bit 3)
Step 10: Set Code ← Code + (NC shifted by 4 bits to left).
Step 11: Set Code ← Code + (L shifted by Shift_L[typ] bits to left).
Step 12: Set Code ← Code + (absM shifted by Shift_M[typ] bits to left).
Step 13: Set Code ← Code + (absN shifted by Shift_N[typ] bits to left).
Step 14: return Code.
Stop.

5.2.3   Applications of DSS with Additional Parameters

The recognition of DSS is one of the fastest and of the most economical methods of encoding two-dimensional geometric objects. It is possible to encode the boundaries in a segmented image as sequences of DSSs, while saving the coordinates of a *single* starting point and an array of economically encoded additional parameters of the DSSs for each boundary. The code requires on average 2.3 bytes per DSS. The parameters *exactly* specify the location of the DSSs in the image thus enabling the *exact* reconstruction of the image. The code is economical. So, for example, when encoding a gray value image with 32 gray levels (Fig. 5.11), a compression rate of 3.1 was reached [9].

The method is amazingly fast: for example, it encodes a binary image of 832×654 pixels containing 57 disk-shaped objects in 20 MS on a PC with a processor of 700 MHz's. Also, the recognition of the *distorted* disks and the estimation of their locations and diameters was performed in that time. Figure 5.12 shows an example of such an image with the recognized disks labeled with blue circles.

Fig. 5.11 Example of an exactly reconstructed image with 32 gray levels whose boundaries of homogeneous regions were encoded with additional DSS parameters.

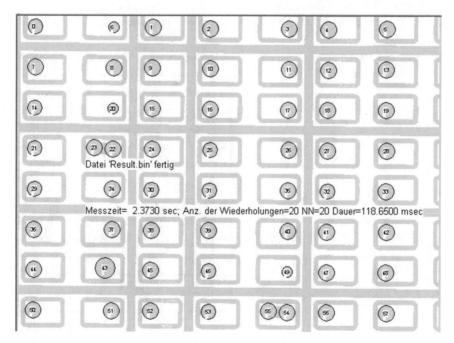

Datei 'Result.bin' fertig

Messzeit= 2.3730 sec; Anz. der Wiederholungen=20 NN=20 Dauer=118.6500 msec

Fig. 5.12 Example of an image of a wafer with recognized disk-shaped objects

Digital straight segments are the best means for estimating the length of digital curves and of the perimeter of a region in 2D images. An experimental comparison of different methods of estimating the length is to be found in [10].

5.3 Estimating the Length of Digital Curves

It is well known that a good estimate of the area of a subset S in a 2D digital image is the number of pixels in S. When increasing the resolution (i.e., making the pixels smaller and multiplying the number of the pixels with the area of a single pixel) the estimate tends to the area of the original continuous image.

It is natural to suppose that the number of cracks in the boundary of a subset could be a good estimate of its perimeter. However, it is not. Consider Fig. 5.13.

Black and red lines represent large pixels while blue lines represent small pixels. The number of pixels and cracks in the black line is $6 \cdot 2 = 12$ pixels and 16 cracks. The length of a black crack is equal to 1. Thus, the perimeter of the black figure is equal to 16.

The red line represents this figure rotated by $45°$. The red line includes 12 pixels and contains 20 cracks of the length equal to 1. The perimeter estimated as the sum of the lengths of the cracks is 20. The number of cracks and the perimeter have changed by 25%. In the case of some other figures the change of the number of cracks due to the rotation can be as large as 41%.

Let us see, how an increased resolution influences the estimates of the area and that of the perimeter. When we increase the resolution, for example, when we use

Fig. 5.13 Example of a rotated figure and of an increased resolution by factor 2

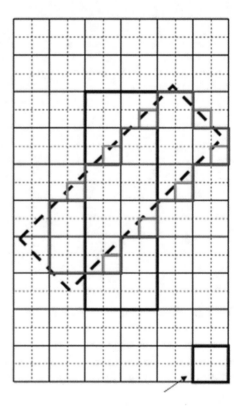

smaller pixels and cracks shown by the blue line in Fig. 5.13, then the number of pixels included by the blue line is 48. However, a blue pixel has an area of 0.25 of the area of a red pixel and the total area of the blue pixels is 48 * 0.25 = 12. It is the same as the total area of the original, not rotated figure. In the case of a figure with a curved boundary the total area of the pixels will tend with increased resolution to the area of the original figure.

A quite different situation takes place in the case of the perimeter. The number of the blue cracks is 40. The length of a blue crack is 0.5 of the length of a red crack. The total length is 40 * 0.5 = 20. It has not changed with the increased resolution!

Thus, the sum of the lengths of the boundary cracks remains at its false value depending on the rotation. This is not a suitable estimate of the perimeter.

A good estimate of the perimeter is the sum of the lengths of the longest DSS in the boundary of the subset. It tends to the perimeter of the original figure at least if the figure is convex or is a polygon whose inner angles are not too small [10].

The length of a DSS is the Euclidean distance between its end points:

$$L = \text{sqrt}((xe - xs)^2 + (ye - ys)^2);$$

where (xs, ys) are the coordinates of the starting and (xe, ye) that of the end point of the DSS. An example is shown in Fig. 5.14.

5.4   Polygonal Approximation

This chapter describes a method of representing curves in two-dimensional digital images as polygon chains. We are using the notion "polygon" for designing a polygon chain or a polygon chain together with its interior.

This representation of a curve is useful for image analysis because the shape of a polygon can be easily investigated by simple geometrical means such as measuring lengths and angles. Polygonal approximation also suggests a new method of estimating the curvature of digital curves. For this purpose, a polygon chain can be replaced by a smooth sequence of circular arcs and straight-line segments. "Smooth" means that each straight segment is the tangent to the previous and to subsequent arc.

5.4.1   The Sector Method

An efficient method called the sector method was suggested in the 70th by Williams [11]. The idea is as follows. Given is a digital curve C. We are looking for a polygon whose points have the distance from the curve C smaller than the predetermined tolerance Eps. Start with some point V_1 of the curve C (Fig. 5.15). Test all subsequent points. For each point V_i with i greater than 1 and distance $D(V_1, V_i)$ from the

Fig. 5.14 Example of estimating the perimeter of a figure with a curved boundary

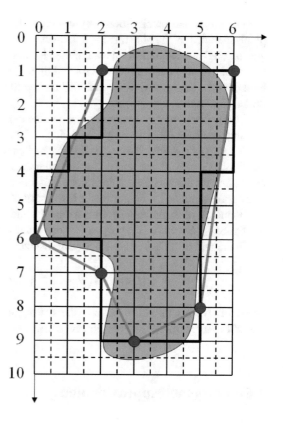

starting point V_1 greater than *Eps*, draw a circle with the radius equal to the tolerance *Eps* and draw two tangents from V_1 to the circle.

The space between the tangents composes a sector, while each straight-line L through V_1 lying in the sector has the property that the distance from the center of the circle to L is less than the tolerance *Eps*. If the next point V_i (V_3 or V_4 in Fig. 5.15) lies in the sector, then a new circle with its center at V_i and a sector S of this circle are constructed. The new sector is the intersection of S with the old one. If the next point V_i (e.g., V_5 in Fig. 5.15) lies outside of the new sector, then there are some points between V_1 and V_i that have a distance to the straight-line L through V_1 and V_i greater than *Eps*.

Therefore, the segment (V_1, V_i) cannot serve as a polygon edge. The point V_{i-1} before V_i is then the last point of the current segment. Segment $[V_1, V_{i-1}]$ is then the edge of the polygon. The construction of the next polygon edge is starting at the point V_{i-1}.

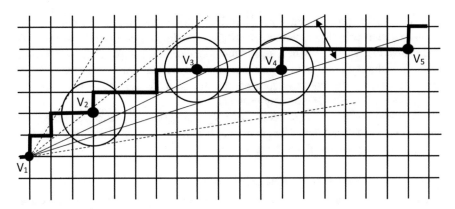

Fig. 5.15 The sector method

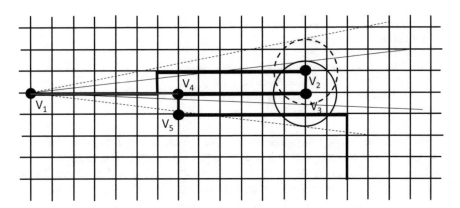

Fig. 5.16 Example of a turning curve

5.4.2 Improvement of the Sector Method

The sector method is fast because it tests each point of the given curve only once. However, it does not guarantee that the polygon is close to the curve if the curve makes a turn and goes back.

Consider please the Fig. 5.16. The actual sector is the intersection of two sectors about the circles with centers at points V_2 and V_3. The sector is shown with thin solid lines. Both points V_2 and V_3 lie in sector. The actual point moves along the curve. It remains in the sector until it reaches the point V_5 which is outside the sector. Then the edge of the polygon will be drawn from the starting point V_1 to point V_4 as shown with the red line. A great part of the curve containing the points V_2 and V_3 is lost.

The following additional condition has been introduced for interrupting the current segment of the curve that should be approximated by the current edge of the polygon. The point V_2 (Fig. 5.16) of the curve which has the greatest distance from the starting point V_1 must be fixed during the processing of the curve. If the distance of the projection of the actual point V_i onto the straight-line segment $[V_1, V_2]$ is less than $|V_1, V_2| - Eps$, where Eps is the tolerance of the approximation, then the point V_2 is the last point of the current segment of the curve to be approximated by the next edge of the polygon and the starting point of the next edge. This test guaranties that no part of a turning curve is lost.

We present below the algorithms "Approx" and "Follow" which read the loops and approximate each loop of length greater than a predefined minimum length by a polygon with a predetermined tolerance "eps".

Algorithm "Approx(minLenght, eps)"

The variables "nPolygon", "nVert", "nLoop", and the arrays "Loop" and "Directions" are global.

Step 1: Declare integer variables first, iLoop, last and length.
Step 2: Initialize variables nPolygon ← 0, iLoop ← 0 and nVert ← 0.
Step 3: Repeat until iLoop is less than nLoop:
    3.1: If iLoop is equal to 0, then set first ← 0.
        else set first ← Loop[iLoop - 1].Last + 1.
    3.2: Set last ← Loop[iLoop].Last.
    3.3: Set Length ← last - first + 1.
    3.4: If Length is greater or equal to minLength, then
        call Follow(Loop, first, last, Directions, eps).
    3.5: Set iLoop ← iLoop + 1.   (End of repetition of step 3).
Stop.

We describe below the algorithm "Follow". Note that the variables "nVert", "nPolygon" and the arrays "Vert" and "Polygon" are global. The variable "Vect" must be transmitted to the algorithm "CheckComb" in such way that "CheckComb" can change the value of "Vect" declared in "Follow". The algorithm uses the array "Step" defined above in Sect. 5.1.4 for the algorithm "TraceLoop".

The algorithm "Follow" uses two small utility algorithms: "CheckComb" and "ParArea". The algorithms "CheckComb" is especially important. We describe it below. The algorithm "ParArea (iv, P)" calculates the area of a parallelogram that spans the points "Vert [iv −2]", "Vert [iv −1]" and "P/2". If this area is zero, then the points are collinear, and the middle point is overwritten by the third one.

Algorithm "Follow(iL, Loop, first, last, Directions, eps)"

The algorithm uses the field Step[4] defined in Section 3.1.

Step 1: Declare integer variables brake, dir, ibyte, iCrack, and StartVert.
Step 2: Declare variables Crack, Point, Pold, Pstand, StartEdge,
StartLine and Vect as two-dimensional vectors.
Step 3: Initialize variables: Pstand ← Loop.Start, Point ← 2*Pstand,
Pold ← Point, StartEdge ← Pstand, StartLine ← Pstand,
StartVert ← nVert, Vert[nVert] ← Pstand, and
Polygon[nPolygon].firstVert ← nVert.
Step 4: Set nVert ← nVert + 1.
Step 5: Call algorithm CheckComb(StartEdge, Pstand, eps, Vect).
Step 6: Initialize variable iByte ← 0.
Step 7: Repeat until iByte is less or equal to last:
7.1: Set dir ← Directions[iByte].
7.2: Set Crack ← Point + Step[dir].
7.3: Set Point ← Crack + Step[dir].
7.4: Set Pstand ← Point/2.
7.5: Set brake ← result of CheckComb(StartEdge, Pstand, eps, Vect).
7.6: Set iCrack ← iCrack + 1.
7.7: If brake is equal to 1, then
Begin
If nVert is greater than StartVert + 1, then set Vert[nVert - 1] ← Pold / 2.
else set Vert[nVert] ← Pold / 2.
End
7.8: If brake is equal to 2, then
Begin
If nVert is greater than StartVert + 1 and algorithm
"ParArea(nVert, Pold)" returns 0, then set Vert[nVert-1] ← Vect.
else set Vert[nVert] ← Vect.
End
7.9: If brake is greater than 0, then
Begin 7.9
If nVert is greater than StartVert + 1 and algorithm
"ParArea(nVert, Pold)" returns 0, then set StartEdge ← Vert[nVert - 1].
else
Begin
Set StartEdge ← Vert[nVert].
If StartEdge.X is greater than 0 or StartEdge.Y is greater than 0, then
set nVert ← nVert + 1.
End
Set brake ← 0.
End 7.9   (End of "If brake is greater than 0")
7.10: Set PointOld ← Point.
7.11: Set iByte ← iByte + 1.   (End of repetition of step 7).
Step 8: Set Polygon[nPolygon].lastVert ← nVert + 1.
Step 9: Set Polygon[nPolygon].closed ← true.
Step 10: Set Polygon[nPolygon].nCrack ← iCrack.
Step 11: If Polygon[nPolygon].lastVert is greater
than Polygon[nPolygon].lastVert, then set nPolygon ← nPolygon + 1.
Step 12: Return 1.
Stop.

The algorithm "CheckComb(StartEdge, Pstand, eps, Vect)" checks whether the point "Pstand" lies inside the sector with the chord "2*eps" and center at "StartEdge" crossed with the previous sector and with the circle with radius "eps" and center at the point furthest from "StartEdge". The variables PosSectX, PosSectY, NegSectX, NegSectY and the two-dimensional vector "Far" are global which is necessary because they should retain their values for multiple calls of "CheckComb". The parameters StartEdge, Pstand, and Vect are vectors. The vector "Vect" should be transferred to "CheckComb" in such a way that it can be changed by "CheckComb" and this change can be used externally.

Algorithm CheckComb(StartEdge, Pstand, eps, iVect2 Vect)

Step 1: Declare variables Length, Sine, Cos, Proj, PosTangX, PosTanfY,
                                        NegTangX, and NegTanfY.
Step 2: Declare variable Line as a two-dimensional vector.
Step 3: If StartEdge is equal Pstand, then
       Begin
        Set PosSectX $\leftarrow$ -1000.0.
        Return 0.
       End
Step 4: Set Line $\leftarrow$ Pstand - StartEdge.
Step 5: Set Length $\leftarrow$ Sqrt(Line.X^2 + Line.Y^2).
Step 6: If Length is less than eps, then return 0.
Step 7: If PosSectX equals -1000.0, then
       Begin
        Set Sine $\leftarrow$ eps / Length.
        Set Cos $\leftarrow$ Sqrt(1.0 - $Sine^2$).
        Set PosSectX $\leftarrow$ Line.X*Cos - LineY*Sine.
        Set PosSectY $\leftarrow$ Line.X*Sine - LineY*Cos.
        Set NegSectX $\leftarrow$ Line.X*Cos + LineY*Sine.
        Set NegSectY $\leftarrow$ Line.X*Sine + LineY*Cos.
        Set Far $\leftarrow$ Pstand.
        return 0.
       End
Step 9: Set Proj $\leftarrow$ (Far.X - StartEdge.X)*(Pstand.X - Far.X) +
                        (Far.Y - StartEdge.Y)*(Pstand.Y - Far.Y).
Step 10: If Proj is less than -eps*Sqrt((Far.X - StartEdge.X)^2 +
                        (Far.Y - StartEdge.Y)^2)), then
       Begin
        Set Vect $\leftarrow$ Far.
        PosSect.X $\leftarrow$ -1000.0.
        return 2.
       End
Step 11: If Proj is greater or equal 0, then set Far $\leftarrow$ Pstand.
Step 12: If PosSectX*Line.Y - PosSectY*Line.X is greater 0 or
      NegSectX*Line.Y - NegSectY*Line.X is less than 0, then
       Begin
        PosSectX - -1000.0.
        return 1.
       End
Step 13: If PosSectX is not equal to -1000.0 and Length is greater than eps, then
       Begin
        Set Sine - eps/Length.
        Set Cos - Sqrt(1.0 - $Sine^2$).
        Set PosTangX $\leftarrow$ Line.X*Cos - LineY*Sine.
        Set PosTangY $\leftarrow$ Line.X*Sine + LineY*Cos.
        Set NegTangX $\leftarrow$ Line.X*Cos + LineY*Sine.
        Set NegTangY $\leftarrow$ -Line.X*Sine + LineY*Cos.

If NegSectX*NegTangY - NegSectY*NegTangX is greater than 0, then
Begin
  Set NegSectX ← NegTangX.
  Set NegSectY ← NegTangY.
End
If PosSectX*PosTangY - PosSectY*PosTangX is greater than 0, then
Begin
  Set PosSectX ← PosTangX.
  Set PosSectY ← PosTangY.
End
    End   (End of If PosSectX is not equal to -1000.0 and...)
Step 14: return 0.
Stop.

Experiments have shown that practically useful polygons can be obtained only if the loops have been obtained from indexed images with a low number of represented colors. In the case of a true color image a great majority of the loops are small loops of four cracks composing the boundary of a single pixel. This is so because in a color image almost each pixel is surrounded by pixels of different colors. In an indexed image the relative number of small loops is much smaller. If one decides to make polygons only for loops containing a minimum number of cracks, then all created polygons will have a reasonable size and most of them can be used for solving practical problems.

Figure 5.17 shows an image indexed with 17 colors while Fig. 5.18 presents the polygons of loops with at least 150 cracks approximated with the tolerance Eps = 2.1.

Fig. 5.17 Indexed image with 17 colors

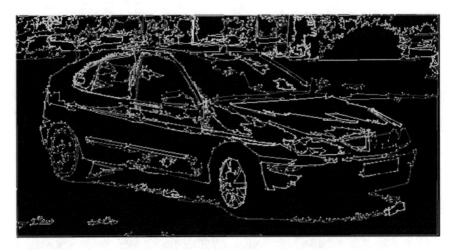

Fig. 5.18 Polygons of loops of the image of Fig. 5.17 with at least 150 cracks

5.4.3 Applications of Polygonal Approximation

One of the possible applications of polygonal approximation is the recognition of objects by analyzing their shape. Polygons obtained by polygonal approximation are sequences of straight segments. It is easy to calculate the lengths of the segments, the angles between two segments and their relative locations. Thus, it becomes possible to analyze the shape of objects represented in the image.

Let us consider the example of recognizing cars in images representing a site of a street in a city. Important features of a car are its wheels. The cap of a wheel looks like a bright circle or ellipse on the dark background of the tire. To find a polygon representing the boundary of a cap of a wheel it is necessary first to check the sizes of polygons. For this purpose, we first calculate for each polygon its box which is a rectangle with sides corresponding to minimum and maximum values of the coordinates of the vertices of the polygon. For a cap of a wheel the size of its box must be limited.

Besides that, the shape of the polygon must satisfy certain limitations. To estimate the shape of a polygon we use the "form factor" or the isoperimetric ratio [12, 13]. The form factor of a two-dimensional subset of a digital image is defined as squared perimeter divided by the area. Both numerator and denominator of the form factor are proportional to the square of the size of the subset. Therefore, the form factor thus defined does not depend on the size of the subset. It depends only on the shape of the subset and is like the "sphericity" used to estimate whether a three-dimensional object is like a sphere.

Our form factor of circle of radius R is equal to $(2\pi R)^2/\pi R^2 = 4\pi \approx 12.6$. It is greater than 4π for all other shapes. It is equal to 16 for a square; 14.95 for an ellipse with the big half-axis equal to two times the small half-axis and 23.75 for an ellipse with the big half-axis equal to four times the small half.

It is rational to consider only polygons with a limited form factor. However, a polygon of a wheel of a car can look like the example shown in Fig. 5.19.

The indentations can make the form factor essentially greater than that of an ellipse. The indentations appear because some spokes or the spaces between spokes have the same color as the tire. The indentations can be removed if we take the convex hull of a polygon. The convex hull of a polygon is the smallest convex set containing the polygon. Many algorithms for calculating the convex hull of a set of points have been published. We are using the Graham scan [14].

After having calculated the convex hull, we calculate its form factor. If it is near 12.6, for example, less than 13, than we calculate the circle nearest to the convex hull by means of the modified method of least squares described in the next Sect. 5.4.4. If, however, the form factor of the convex hull is greater than 13 and less than 20, then we use the classical method of least squares for finding the ellipse close to the convex hull (Sect. 5.4.5) for which the sum of the squared distances of the vertices of the polygon from the ellipse is minimal.

After having recognized the wheels, we look for an almost horizontal polygon edge lying between the wheels. Such an edge represents the bottom part of the body. It is well visible in the most cases.

The next important features of a car are the frames of the windows. Almost each car has windows whose frames contain an almost horizontal line orthogonal to some almost vertical line near the horizontal line. These lines compose an angle typical for a window.

Also, the shape and the location of the boundary of the roof is an important feature of car. In the most cases there is an arc lying in a certain distance above the wheels. Its center of curvature should lie near or below the bottom of the car almost between the wheels. Also, an almost horizontal line is typical for the roof.

The presence of the most of these features is an indication of a car. Figure 5.20 shows an example of a photograph of a car with signed features.

Polygons are also useful for the recognition of circular or elliptical objects. Both the modified method of least squares and the classical method of least squares used for finding a circle or an ellipse close to a given set of points are very robust: They

Fig. 5.20 A car with some recognized features

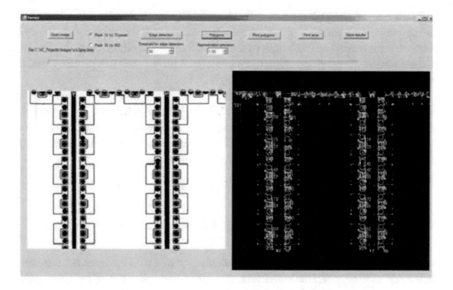

Fig. 5.21 Example of an image with recognized circles labeled with red lines

find the optimal circle or the optimal ellipse even if the points of the given set are strongly distorted by noise. They work even in cases when the given set of points represents only a small part of the curve. We have used the modified method for example for checking the quality of solder bumps in wafers in a factory producing electronic devices (Fig. 5.21).

Some circles where strongly distorted and disintegrated in short arcs, as shown in Fig. 5.22. Such circles have been nevertheless recognized and their parameters as the coordinates of the center and the radius where correctly estimated.

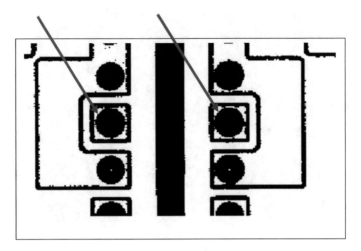

Fig. 5.22 Examples of correctly recognized distorted circles

In the case of recognizing a car we calculate ellipses or circles approximating the curves probably representing the wheels. We consider the relation of the distance between their centers to the length of the major axis of the ellipse or to the diameter of the circle. If the relation lies between predefined limits, then it is probable that the curves represent the wheels of a car.

To make the recognition of a car safer, additional features are to be investigated. Such a feature may be, for example, a series of nearly horizontal segments wide above the wheels and lying near a horizontal line. It would correspond to the roof of the car. It can also be an approximately rectangular polygon between the wheels and the roof. Such a polygon would correspond to a window of a car.

5.4.4   Algorithm for Recognizing Circles in Distorted Images

The method is like the classical method of least squares with the essential difference that we do minimize not the sum of the squares of the deviations of the given points from the sought for circle but rather the sum of the squares of the differences of the area of a circle with the unknown center (x_c, y_c) through the given point (x_i, y_i) from the area of a circle with the unknown radius R:

$$\text{Deviat} = \sum (R^2 - (x_i - x_c)^2 - (y_i - y_c)^2\ )^2. \tag{5.2}$$

The sum is calculated according to the index i of the points, where i gets N values from 0 to $N - 1$. This criterium differs from the classical one:

$$\text{Crit} = \sum_i \left(R - \sqrt{(x_i - x_c)^2 - (y_i - y_c)^2} \right)^2; \tag{5.3}$$

in that instead of the difference "(distance of a point from the middle point) — radius" we use the difference "distance^2 — radius^2". This makes the solution much easier and faster while the difference of the found circle from the classical one is exceedingly small.

Let us derive the necessary equations. We take the partial derivatives of the expression (5.2) after x_c, y_c and R.

$$\partial \text{Deviat}/\partial x_c = 4 * \sum \left(R^2 - (x_i - x_c)^2 - (y_i - y_c)^2 \right) * (x_i - x_c); \qquad (5.4)$$

$$\partial \text{Deviat}/\partial y_c = 4 * \sum \left(R^2 - (x_i - x_c)^2 - (y_i - y_c)^2 \right) * (y_i - y_c); \qquad (5.5)$$

$$\partial \text{Deviat}/\partial R = 4 * \sum \left(R^2 - (x_i - x_c)^2 - (y_i - y_c)^2 \right) * R; \qquad (5.6)$$

and set them equal to 0. We obtain three equations:

$$\sum \left(R^2 - (x_i - x_c)^2 - (y_i - y_c)^2 \right) * (x_i - x_c) = 0; \qquad (5.7)$$

$$\sum \left(R^2 - (x_i - x_c)^2 - (y_i - y_c)^2 \right) * (y_i - y_c) = 0; \qquad (5.8)$$

$$\sum \left(R^2 - (x_i - x_c)^2 - (y_i - y_c)^2 \right) = 0; \qquad (5.9)$$

The left parts are polynomials containing the variables x_c, y_c and R with powers 1, 2 and 3. We demonstrate in what follows how to transform Eqs. (5.7) and (5.8) to linear ones.

We open the round brackets in (5.9) and obtain:

$$\sum (R^2 - x_i^2 + 2 * x_i * x_c - x_c^2 - y_i^2 + 2 * y_i * y_c - y_c^2) = 0;$$

or

$$\sum (R^2 - x_c^2 - y_c^2) = \sum (x_i^2 - 2 * x_c * x_i + y_i^2 - 2 * y_c * y_i).$$

or

$$N * (R^2 - x_c^2 - y_c^2) = \sum (x_i^2 - 2 * x_c * x_i + y_i^2 - 2 * y_c * y_i). \qquad (5.10)$$

since $(R^2 - x_c^2 - y_c^2)$ does not depend on i.

Let us now transform the Eq. (5.7) to separate the expression $R^2 - x_c^2 - y_c^2$.

$$\sum \left(R^2 - (x_i - x_c)^2 - (y_i - y_c)^2 \right) * (x_i - x_c) =$$

$$\sum \left(R^2 - (x_i - x_c)^2 - (y_i - y_c)^2 \right) * x_i - \sum \left(R^2 - (x_i - x_c)^2 - (y_i - y_c)^2 \right) * x_c =$$

$$\sum \left(R^2 - (x_i - x_c)^2 - (y_i - y_c)^2 \right) * x_i - x_c * \sum \left(R^2 - (x_i - x_c)^2 - (y_i - y_c)^2 \right) = 0$$

The second term $x_c * \sum(\ldots)$ is according to (5.9) equal to 0. Thus:

$$\sum \left(R^2 - (x_i - x_c)^2 - (y_i - y_c)^2 \right) * x_i =$$
$$\left(R^2 - x_c^2 - y_c^2 \right) * \sum x_i - \sum (x_i^2 - 2 * x_i * x_c + y_i^2 - 2 * y_i * y_c) * x_i = 0$$

According to (5.10) we replace $\left(R^2 - x_c^2 - y_c^2 \right)$ with $\sum (x_i^2 - 2 * x_c * x_i + y_i^2 - 2 * y_c * y_i)/N$ in the first term and perform the multiplication with x_i in the second one:

$$\sum x_i * \sum (x_i^2 - 2 * x_c * x_i + y_i^2 - 2 * y_c * y_i)/N$$
$$- \sum (x_i^3 - 2 * x_i^2 * x_c + x_i * y_i^2 - 2 * y_i * y_c * x_i)$$
$$= 0$$

Let us denote each sum $\sum x_i^m * y_i^n$ with S_{mn}. Then:

$$S_{10} * (S_{20} - 2 * x_c * S_{10} + S_{02} * S_{01} - 2 * y_c * S_{01})$$
$$- (S_{30} + 2 * x_c * S_{20} + S_{12} + 2 * y_c * S_{11}) * N$$
$$= 0.$$

or

$$2 * x_c \left(N * S_{20} - S_{10}{}^2 \right) + 2 * y_c * (N * S_{11} - S_{10} * S_{01})$$
$$= (S_{30} + S_{12}) * N - S_{10} * S_{20} - S_{02} * S_{01} \tag{5.11}$$

This is a linear equation in x_c and y_c. After similar transformations, the Eq. (5.8) also becomes a linear equation:

$$2 * x_c (N * S_{11} - S_{10} * S_{01}) + 2 * y_c * (N * S_{02} - S_{02}^2)$$
$$= (S_{03} - S_{21}) * N - S_{01} * S_{20} - S_{02} * S_{01} \tag{5.12}$$

Equations (5.11) and (5.12) can be easily solved. The solutions can be set in the Eq. (5.10) rewritten in the S_{mn} notation:

$$R^2 - x_c^2 - y_c^2 = (S_{20} - 2 * x_c * S_{10} + S_{02} - 2 * y_c * S_{01})/N \tag{5.13}$$

to obtain the value of R.

We present below the algorithm "MinAreaN2" performing these solutions. The parameters "radius", "x0" and "y0" must be transmitted in such a way that they can be changed by the algorithm. The algorithm calculates the estimates "x0" and "y0" of the coordinates of the center and the estimate "radius" of the radius of the optimal circle with the minimum deviation from the given set "P[np]" of points. The found values and the 13 sums used for the calculation are assigned to the arc with the index "ia".

Algorithm MinAreaN2(ia, vector field P, Start, np, radius, x0, y0)

Step 1: Declare the variables SumX, SumY, SumX2, SumY2, SumXY, SumX3, SumY3,
SumX2Y, SumXY2, SumX4, SumX2Y2, SumY4,
a1, a2, b1, b2, c1, c2, Crit, det, fx, fy, mx, my, N, R2, and ip.

Step 2: Set N ← np.

Step 3: Set the value of the variables SumX, SumY, SumX2, SumY2, SumXY, SumX3,
SumY3, SumX2Y, SumXY2, SumX4, SumX2Y2, SumY4 to zero.

Step 4: Set ip ← Start.

Step 5: Repeat until ip is less than Start + np: (Loop over the set of points)
Begin
  Set fx ← P[ip].X.
  Set fy ← P[ip].Y.
  Set SumX ← SumX + fx.
  Set SumY ← SumY + fy.
  Set SumX2 ← SumX2 + fx * fx.
  Set SumY2 ← SumY2 + fy * fy.
  Set SumXY ← SumXY + fx * fy.
  Set SumX3 ← SumX3 + fx * fx * fx.
  Set SumY3 ← SumY3 + fy * fy * fy.
  Set SumX2Y ← SumX2Y + fx * fx * fy.
  Set SumXY2 ← SumXY2 + fx * fy * fy.
  Set SumX4 ← SumX4 + fx * fx * fx * fx.
  Set SumX2Y2 ← SumX2Y2 + fx * fx * fy * fy.
  Set SumY4 ← SumY4 + fy * fy * fy * fy.
  Set ip ← ip + 1.
End   (End repetition Step 5)

Step 6: Set the following 7 variables:
a1 ← 2 * (SumX * SumX - N * SumX2).
b1 ← 2 * (SumX * SumY - N * SumXY).
a2 ← 2 * (SumX * SumY - N * SumXY).
b2 ← 2 * (SumY * SumY - N * SumY2).
c1 ← SumX2 * SumX - N * SumX3 + SumX * SumY2 - N * SumXY2.
c2 ← SumX2 * SumY - N * SumY3 + SumY * SumY2 - N * SumX2Y.
det ← a1 * b2 - a2 * b1.

Step 7: If |det| is less than 0.00001, then return -1.0.

Step 8: Set the following 3 variables:
mx ← (c1 * b2 - c2 * b1) / det.
my ← (a1 * c2 - a2 * c1) / det.
R2 ← (SumX2 - 2 * SumX * mx - 2 * SumY *my + SumY2) / N + mx*mx + my*my.

Step 9: If R2 is less or equal to 0, then return -1.0.

Step 10: Set the following 4 variables:
x0 ← mx.
y0 ← my.
radius ← Sqrt(R2).
Crit ← 0.0.

Step 11: Set ip ← Start.

Step 12: Repeat until ip is less than Start + np:

Begin
  fx ← P[ip].X.
  fy ← P[ip].Y.
  Crit ← Crit + (radius - Sqrt((fx - mx) * (fx - mx) + (fy - my)* (fy - my)))^2.
  ip ← ip + 1.
 End.   (End of repetition Step 12.)
Step 13: return Sqrt(Crit / np).
Stop   (End of the algorithm MinAreaN2.)

Examples of recognized circles are shown in Fig. 5.21 above.

5.4.5   Recognition of Ellipses in Distorted Images

Due to the good properties of our method for circle recognition we came to the idea to use this method to recognize the wheels of the bicycles which are ideal circles. However, if the bike is positioned so that the plane of its frame makes an acute angle with the viewing direction, then the wheels look like ellipses, rather than like circles. Therefore, we also need a method of recognizing ellipses. Unfortunately, we have not succeeded with generalizing our method of circle recognition (Sect. 5.4.4) for ellipses. We have tried to use the well-known method of conjugate gradients. However, our experiments have shown that this method is not robust: Sometimes it fails when the points to which the ellipse is to be fitted do not lie near an ellipse.

Since an ellipse is defined by only a small number of parameters, namely by five, it is possible to use the classical procedure of least squares as described in the next section.

5.4.6   Mathematical Foundation of Ellipse Recognition

An ellipse with axes parallel to the coordinate axes of a Cartesian coordinate system and with the center lying in the origin has the well-known equation:

$$x^2/a^2 + y^2/b^2 = 1;$$

However, we need considering the general case of a shifted and inclined ellipse. We use the general equation of a conic section:

$$Ax^2 + Bxy + Cy^2 + Dx + Ey + F = 0;$$

Our aim is to find parameters of an ellipse for which the sum of the squared distances of a set of given points from the ellipse is minimal. Thus, our objective function is

$$f = \sum_i \left(Ax_i^2 + Bx_i y_i + Cy_i^2 + Dx_i + Ey_i + F \right)^2. \tag{5.14}$$

The expression in the parentheses is approximately proportional to the distance of the point (x_i, y_i) from the ellipse. It contains 6 unknown coefficients A, B, C, D, E and F. However, it is well known that an ellipse is uniquely defined by 5 parameters. Therefore, we divide all terms of (5.14) by A and denote the new coefficients as follows:

$$B/A = 2k_1;$$
$$C/A = k_2;$$
$$D/A = 2k_3;$$
$$E/A = 2k_4;$$
$$F/A = k_5;$$

The transformed objective function is:

$$f = \sum_i \left(x_i^2 + 2k_1 x_i y_i + k_2 y_i^2 + 2k_3 x_i + 2k_4 y_i + k_5 \right)^2. \tag{5.15}$$

The partial derivative of f to k_1 is:

$$\frac{\partial f}{\partial k_1} = 4 \sum_i \left(x_i^2 + 2k_1 x_i y_i + k_2 y_i^2 + 2k_3 x_i + 2k_4 y_i + k_5 \right) * x_i y_i.$$

After multiplying all terms in the parenthesis with $x_i * y_i$ we obtain:

$$\frac{\partial f}{\partial k_1} = 4\sum_i \left(x_i^3 y_i + 2k_1 x_i^2 y_i^2 + k_2 x_i y_i^3 + 2k_3 x_i^2 y_i + 2k_4 x_i y_i^2 + k_5 x_i y_i\right)$$

We divide it with 4, denote each $\sum_i x_i^m y_i^n$ by $S(m, n)$ and obtain:

$$\frac{\partial f}{\partial k_1} = S(3, 1) + 2k_1 S(2, 2) + k_2 S(1, 3) + 2k_3 S(2, 1) + 2k_4 S(1, 2) + k_5 S(1, 1)$$

By setting it equal to 0 we obtain the first of the five equations for the unknowns k_1, k_2, k_3, k_4 and k_5.

$$2k_1 S(2, 2) + k_2 S(1, 3) + 2k_3 S(2, 1) + 2k_4 S(1, 2) + k_5 S(1, 1) = -S(3, 1)$$

In a similar way we obtain the other four equations:

$$2k_1 S(1, 3) + k_2 S(0, 4) + 2k_3 S(1, 2) + 2k_4 S(0, 3) + k_5 S(0, 2) = -S(2, 2)$$
$$2k_1 S(2, 1) + k_2 S(1, 2) + 2k_3 S(2, 0) + 2k_4 S(1, 1) + k_5 S(1, 0) = -S(3, 0)$$
$$2k_1 S(1, 2) + k_2 S(0, 3) + 2k_3 S(1, 1) + 2k_4 S(0, 2) + k_5 S(0, 1) = -S(2, 1)$$
$$2k_1 S(1, 1) + k_2 S(0, 2) + 2k_3 S(1, 0) + 2k_4 S(0, 1) + k_5 S(0, 0) = -S(2, 0)$$

This system of equations is solved by the well-known Gauss method implemented in the algorithm "GetEllipse" using the well-known method "Gauss". The sums "$S(x^m, y^n)$" are calculated by the utility algorithm "MakeSums".

5.4.7   Algorithm of Recognizing an Ellipse

Below is the algorithm "GetEllipse". It obtains as parameters the field "Vert" of two-dimensional vectors and two integer variables "iv" and "nPoints" indicating that the algorithm shall use "nPoints" vectors "Vert" while starting at "Vert[iv]". The algorithm calculates the parameters "Delta", "f", "a", "b", "c", and "d" of the ellipse having the minimum sum of squared distances from the given "nPoints" vectors. These parameters should be defined as global.

Algorithm "GetEllipse(field Vert, integer iv, integer nPonts)

Step 1: Define two two-dimensional fields A[5, 5] and B[5, 1].
Step 2: Define the field Sum of 15 elements.
Step 3: Call the utility algorithm MakeSums(Vert, iv, nPoints, Sum).   (Calculates 15 sums)
Step 4: Initialize the 25 elements of the field A:
  Begin
    A[0, 0] ← 2.0 * Sum[12];
    A[0, 1] ← Sum[11];
    A[0, 2] ← 2.0 * Sum[8];
    A[0, 3] ← 2.0 * Sum[7];
    A[0, 4] ← Sum[4];
    A[1, 0] ← 2.0 * Sum[11];
    A[1, 1] ← Sum[10];
    A[1, 2] ← 2.0 * Sum[7];
    A[1, 3] ← 2.0 * Sum[6];
    A[1, 4] ← Sum[3];
    A[2, 0] ← 2.0 * Sum[8];
    A[2, 1] ← Sum[7];
    A[2, 2] ← 2.0 * Sum[5];
    A[2, 3] ← 2.0 * Sum[4];
    A[2, 4] ← Sum[2];
    A[3, 0] ← 2.0 * Sum[7];
    A[3, 1] ← Sum[6];
    A[3, 2] ← 2.0 * Sum[4];
    A[3, 3] ← 2.0 * Sum[3];
    A[3, 4] ← Sum[1];
    A[4, 0] ← 2.0 * Sum[4];
    A[4, 1] ← Sum[3];
    A[4, 2] ← 2.0 * Sum[2];
    A[4, 3] ← 2.0 * Sum[1];
    A[4, 4] ← Sum[0];
  End
Step 5: Initialize the 5 elements of the field B:
  Begin
    B[0, 0] ← -Sum[13];
    B[1, 0] ← -Sum[12];
    B[2, 0] ← -Sum[9];
    B[3, 0] ← -Sum[8];
    B[4, 0] ← -Sum[5];
  End
Step 6: Call the standard algorithm Gauss(A, 5, B, 1).   (Solution is placed in "B")
Step 7: Define utility variables BigDelta, S, a2, b2, aprim, cprim.
Step 7: Calculate the results:
  Begin
    f ← -0.5 * Atan2(2.0 * B[0, 0], 1.0 - B[1, 0]);
    c ← (B[0, 0] * B[3, 0] - B[1, 0] * B[2, 0]) / (B[1, 0] - B[0, 0] * B[0, 0]);
    d ← (B[0, 0] * B[2, 0] - B[3, 0]) / (B[1, 0] - B[0, 0] * B[0, 0]);
    Delta ← B[1, 0] - B[0, 0] * B[0, 0];

BigDelta ← B[1, 0] * B[4, 0] + B[0, 0] * B[3, 0] * B[2, 0] + B[0, 0] * B[3, 0] * B[2, 0] -
            B[2, 0] * B[1, 0] * B[2, 0] - B[3, 0] * B[3, 0] - B[4, 0] * B[0, 0] * B[0, 0];
S ← 1.0 + B[1, 0];
aprim ← (1.0 + B[1, 0] + Sqrt((1.0 - B[1, 0]) * (1.0 - B[1, 0]) +
                                        4.0 * B[0, 0] * B[0, 0])) * 0.5;
cprim ← (1.0 + B[1, 0] - Sqrt((1.0 - B[1, 0]) * (1.0 - B[1, 0]) +
                                        4.0 * B[0, 0] * B[0, 0])) * 0.5;
a2 ← -BigDelta / aprim / Delta;
b2 ← -BigDelta / cprim / Delta;
a ← Sqrt(a2);
b ← Sqrt(b2);
End
Step 8: If Delta is positive, then return 1.   (The found curve is an ellipse)
        else return -1.   (The found curve is no ellipse)
Stop

Now we describe the utility algorithm "Make Sums":

Utility algorithm MakeSums(field Vert, integer iv, integer nPoints, field Sum)

Step 1: Define the variables I, X, and Y.
Step 2: Initialize the 15 elements of Sum with 0.
Step 3: Set I ← iv.
Step 4: Repeat until I is less than iv + nPoints:
  Begin
    Set X ← Vert[I].X.
    Set Y ← Vert[I].Y.
    Set Sum[0] ← Sum[0] + 1.
    Set Sum[1] ← Sum[1] + Y.
    Set Sum[2] ← Sum[2] + X.
    Set Sum[3] ← Sum[3] + Y*Y.
    Set Sum[4] ← Sum[4] + X*Y.
    Set Sum[5] ← Sum[5] + X*X.
    Set Sum[6] ← Sum[6] + Y*Y*Y.
    Set Sum[7] ← Sum[7] + X*Y*Y.
    Set Sum[8] ← Sum[8] + X*X*Y.
    Set Sum[9] ← Sum[9] + X*X*X.
    Set Sum[10] ← Sum[10] + Y*Y*Y*Y.
    Set Sum[11] ← Sum[11] + X*Y*Y*Y.
    Set Sum[12] ← Sum[12] + X*X*Y*Y.
    Set Sum[13] ← Sum[13] + X*X*X*Y.
    Set Sum[14] ← Sum[14] + X*X*X*X.
    Set I] ← I + 1;
  End   (End of step 4).
  Stop

Figure 5.23 shows an example of an ellipse recognized from the polygons
calculated in a project processing images of bicycles.

Fig. 5.23 Example of an ellipse recognized from the polygons calculated in a project processing images of bicycles

References

1. Debled-Rennesson I, Reveillès J-P. A linear algorithm for segmentation of digital curves. Int J Pattern Recognit Artific Intell. 1995;9:635–62.
2. Debled-Rennesson I. Etude et reconnaissance des droites et plans discrets. PhD thesis, Université Louis Pasteur, Strasbourg. 1995.
3. Anderson TA, Kim CE. Representation of digital line segments and their preimages. Comp Vis Graph Image Process. 1985;30:279–88.
4. Freeman H. Computer processing of line-drawing images. Comput Surv. 1974;6:57–97.
5. Reveilles JP. Géométrie discrète, calcul en nombres entier et algorithmique. Thèse d'état, Université Louis Pasteur, Strasbourg. 1991.
6. Rosenfeld A, Kak AC. Digital picture processing. Academic; 1976.
7. Reveilles JP. Structure des Droit Discretes, Journée mathématique et informatique. 1989.
8. Freeman H. On the encoding of arbitrary geometric configurations. IRE TransElectronComput. 1961;EC-10:260–8.
9. Kovalevsky V. Application of digital straight segments to economical image encoding. In: Ahronovitz E, Fiorio C, editors. DGCI, LNSC; 1568. p. 118–35.
10. Klette R, Kovalevsky V, Yip B. On length estimation of digital curves. In: Part of the SPIE conference on vision geometry VIII, Denver/Colorado, vol. 3811. SPIE; 1999. p. 117–28.
11. Williams C. An efficient algorithm for the piecewise linear approximation of planar curves. Comp Graph Image Process. 1978;8:286–93.

12. Wadell H. Volume, Shape and roundness of quartz particles. J Geol. 1935;43(3):250–80. https://doi.org/10.1086/624298.
13. Berger M. Geometry revealed: a Jacob's Ladder to modern higher geometry. Springer; 2010. p. 295–6. ISBN 9783540709978
14. Graham RL. An efficient algorithm for determining the convex hull of a finite planar set (PDF). Inf Process Lett. 1972;1(4):132–3. https://doi.org/10.1016/0020-0190(72)90045-2.
15. Kovalevsky V. Geometry of locally finite spaces. Editing House Dr. Baerbel Kovalevski; 2008. ISBN 978-3-981 2252-0-4.

Chapter 6
Edge Detection in 2D Images

Abstract This chapter describes first the preprocessing of the image necessary for a successful edge detection. It consists of noise reduction with the Sigma filter and of sharpening the edges with the extreme filter. A new algorithm for edge detection is described. It is faster and more precise than the known algorithms.

Keywords Edge detection · Preprocessing · Drawbacks of Median · Sigma filter · Extreme value filter · New method of edge detection · Color difference · Comparison with Canny edge detection · Edges in color images · Applications of edge detection · Recognition of bicycles

Efficient edge detection needs certain preprocessing of the image: reducing the Gaussian noise and sharpening the ramps, which is necessary because of the following reasons.

The brightness at the boundary of a homogeneous region changes sometimes not abruptly but gradually. In a color or an indexed image with many different indices the brightness near a boundary can change at many parallel lines near the boundary as shown in Fig. 6.1.

Consider please the following example of a small fragment of an image (Fig. 6.2).

Parallel lines occur near the boundary of a homogeneous region if the color slightly changes there. Such lines give a false information of the location and the shape of the boundary. The job of preprocessing is to remove these lines.

6.1 Important Preprocessing

Two algorithms of preprocessing the images are important for successive edge detection: suppressing Gaussian noise and sharpening the boundaries. This preprocessing is especially important both for the edge detection algorithm described in [1] and for the new algorithm.

© The Author(s), under exclusive license to Springer Nature Singapore Pte Ltd. 2021   113
V. Kovalevsky, *Image Processing with Cellular Topology*,
https://doi.org/10.1007/978-981-16-5772-6_6

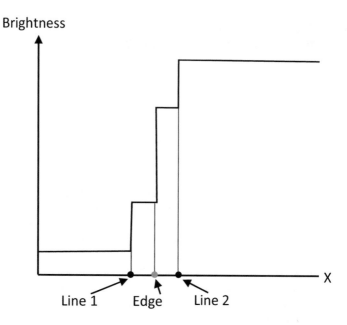

Fig. 6.1 Parallel lines and the edge at the boundary of a homogeneous region

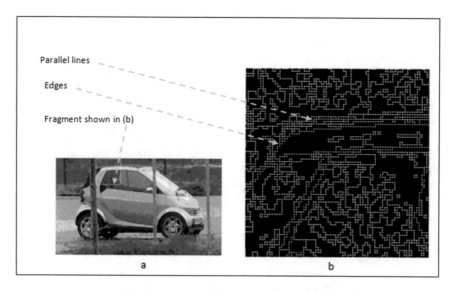

Fig. 6.2 A car image indexed with 90 colors (**a**) and a fragment with lines (green) and edges (red) (**b**)

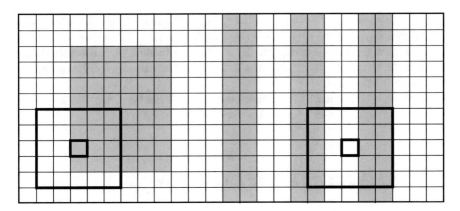

Fig. 6.3 Original image and the gliding window of 5 × 5 pixels

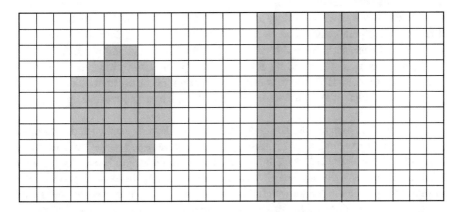

Fig. 6.4 The image of Fig. 6.3 after filtering with median of 5 × 5 pixels

It is well known that the averaging of the color channels provides the most efficient suppression of Gaussian noise. The averaging filter whose gliding window has the width of $W = 2h + 1$ pixel, reduces the standard value of the noise by the factor equal to W. However, it transforms steep edges of homogeneous regions to ramps of the width W. The edges are strongly blurred. Therefore, it cannot be used.

Most textbooks on image processing recommend using median filters for noise suppression. A median filter sorts the intensities of colors in the gliding window and replaces the intensity in the middle of the gliding window by the intensity staying in the middle of the sorted sequence. However, almost no textbook draws the attention of the reader to especially important drawbacks of the median filter: It heavily distorts the image. So a median filter with the gliding window of $(2 * h + 1)^2$ pixels (Fig. 6.3) deletes each stripe of the width of less than or equal to h pixels. It also deletes a triangular part of approximately $2\,h$ pixels at each corner of a rectangular shape (Fig. 6.4). Even more, it inverts a part of the image containing some parallel stripes of the width h if the width of the spaces between the stripes is also equal to h (compare Figs. 6.3 and 6.4). This is easily understandable if the reader notices that median takes decisions according to the majority: the central pixel becomes dark if the most pixels in the gliding window are dark, and it becomes light if the majority of pixels are light.

Because of these reasons, the median filter shall not be used. We suggest an efficient method in the chapter below.

6.1.1   Sigma Filter: The Most Efficient One

The sigma filter reduces the noise in the same way as the averaging filter: by averaging many gray values or colors. The idea of the sigma-filter consists in averaging only those intensities, i.e., gray values or intensities of color channels, in a gliding window which differ from the intensity of the central pixel by no more than by a fixed parameter called "tolerance". According to this idea, the sigma filter reduces the Gaussian noise and retains the edges in the image not blurred.

Sigma filter was suggested by John-Sen Lee [2] in 1983. However, it remains almost unknown until recently: it has been mentioned in no textbook for image processing known to the author. It was mentioned in a professional paper only once [3].

A filter like the sigma filter has been suggested in 1998 by Tomasi and Manduchi [4] which they have called the *bilateral filter*. They have suggested assigning two kinds of weights to the colors being averaged: a "domain weight" becoming smaller with the increasing distance of the averaged pixel from the central pixel of the gliding window and a "range weight" becoming smaller with increasing difference between the intensities of the colors of the pixel being averaged and that of the central pixel. Both weights can be defined as densities of the Gauss distribution. The filter works well: it reduces the Gaussian noise and preserves the sharpness of the edges. However, it is essentially slower that the sigma filter: e.g., the bilateral filter needs 30 s to process a color image of 2500×3500 pixels, while the simplest sigma filter needs only 7 s. Thus, bilateral filter is approximately four times slower than the sigma filter. The authors of the bilateral filter did not mention the sigma filter among the references.

To explain why the sigma filter works so well we remark that it can be regarded as the first iteration of the well-known EM-algorithm or expectation-maximization algorithm [5]. The subdivision of the colors in the gliding window into two subsets, those being close to the color of the central pixel and those being far away from it, can be considered as the expectation step, while the averaging of the close colors as the maximization step. It is well known that the EM-algorithm converges rather quickly. Therefore, already the single first iteration brings a result which is close to the mean value of the subset of pixels in the gliding window whose colors are close to the color of the middle pixel with a high probability.

When comparing the sigma filter with the bilateral filter, it is possible to remark that sigma filter uses an algorithm like that of the bilateral filter where Gaussian distributions are replaced by simpler "rectangular" distributions which can be calculated much faster.

Let us first present the algorithm for filtering gray-scale images. We denote with "Image" the two-dimensional field representing the input image and with "Output" the global output image. The value "hWind" is the half-width of the gliding window. The width and the height of this window are equal to 2*hWind +1. These values are always odd to make the central pixel of the window well defined. The value "Tol" is the tolerance defining the maximum admissible deviation of the averaged values from the gray value of the central pixel. "Max(a, b)" and "Min(a, b)" are functions defining the maximum or the minimum value of the two arguments.

Algorithm SigmaGray(Image[Width, Height], integer hWind, integer Tol)

Step 1: Define the variables CenterV, MaxDif, MinDif, xEnd,
                                            xStart, yEnd, yStart, Sum, NumSum, x, X, y, and Y.
Step 2: Set Y ← 0,
Step 3: Repeat until Y is less than Height:
    3.1: Set three variables: yStart ← Max(0, Y – hWind),
                              EndJ ← Min(Height – 1, Y + hWind), and X ← 0.
    3.2: Repeat until X is less than Width:
        3.2.1: Set xStart ← Max(0, X – hWind) and xEnd ← Min(Width – 1, X + hWind).
        3.2.2: Set five variables: CenterV ← Image[X, Y] and Sum ← 0, NumSum ← 0,
            MinDif ← Max(0, CV – Tol) and MaxDif ← Min(255, CV + Tol).
        3.2.3: Set y ← yStart.
        3.2.4: Repeat until y is less or equal to yEnd:
            3.2.4.1: Set x ← xStart.
            3.2.4.2: Repeat until x is less or equal to xEnd:
            Begin
              If Image[x, y] is between MinDif and MaxDif, then
                set Sum ← Sum + Image[x, y] and NumSum ← NumSum + 1.
              Set x ← x + 1.
            End   (End of Repeat 3.2.4.2)
            3.2.4.3: Set y ← y + 1.   (End of Repeat 3.2.4)
        3.2.5: If NumSum is positive, then set Output[X, Y] ← Sum/NumSum.
            else set Output[X, Y] ← Image[X, Y].
        3.2.6: Set X ← X + 1.   (End of Repeat 3.2)
    3.3: Set Y ← Y + 1.   (End of Repeat 3)
Stop

Now let us present the simplest algorithm for color images. Some variables are now replaced by small fields of three elements corresponding to three color channels Red, Green, and Blue. The input image "Image" and the global output image "Output" are now three-dimensional fields because each pixel (X, Y) contains three color channels Red, Green, and Blue.

Algorithm SigmaSimpleColor(Image[Width, Height, 3], hWind, Tol)

Step 1: Define the variables c, xEnd, yEnd, I, J, xStart, yStart, X, and Y.
Step 2: Define the fields Color[3], MaxDif[3], MinDif[3], Sum[3], and NumSum[3].
Step 3: Set Y ← 0.
Step 4: Repeat until Y is less than Height:
    4.1: Set three variables: yStart ← Max(Y – hWind, 0),
                                      yEnd ← Min(Y + hWind, Heigh - 1) and X ← 0.
    4.2: Repeat until X is less than Width:
    4.2.1: Set xStart ← Max(X – hWind, 0) and xEnd ← Min(X + hWind, Width – 1).
    4.2.2: Set c ← 0.
    4.2.3: Repeat until c is less than 3:
    4.2.3.1: Set five variables: Color[c] ← Image[X,Y,c], Sum[c] ← 0, NumSum[c] ← 0,
              MinDif[c] ← Max(0,Color[c]–Tol) and MaxDif ← Min(255,Color[c]+Tol).
    4.2.3.2: Set c ← c + 1.   (End Repeat 4.2.3)
    4.2.4: Set J ← yStart.
    4.2.5: Repeat until J is less or equal to yEnd:
    4.2.5.1: Set I ← xStart.
    4.2.5.2: Repeat until I is less or equal to xEnd:
    4.2.5.2.1: Set c ← 0.
    4.2.5.2.2: Repeat until c is less than 3:
      Begin
       If Image[X+I, Y+J, c] is between MinDif[c] and MaxDif[c], then
       set Sum[c] ← Sum[c]+Image[X+I,Y+J,c] and NumSum[c] ← NumSum[c] + 1.
       Set c = c + 1.
      End   (End Repeat 4.2.5.2.2)
    4.2.5.2.3: Set I ← I + 1.   (End of Repeat 4.2.5.2)
    4.2.5.3: Set J ← J + 1.   (End of Repeat 4.2.5)
    4.2.6: Set c ← 0.
    4.2.7: Repeat until c is less than 3:
    4.2.7.1: If NumSum[c] is positive, then set Output[X,Y,c] ← Sum[c]/NumSum[c],
              else set Output[X,Y,c] ← Image[X,Y,c].
    4.2.7.2: Set c ← c + 1.   (End of Repeat 4.2.7)
    4.2.8: Set X ← X + 1.   (End of Repeat 4.2)
    4.3: Set Y ← Y + 1.   (End of Repeat 4)
Stop

This solution works well, but it is rather slow if the size of the gliding window is greater than 5×5 pixels: it needs approximately OPP $= 4*W^2$ operations per pixel for gray images or $9*W^2$ for color images. Let us remark, that in the most practical cases it suffices to use the Sigma filter with the window size of 3×3 or 5×5 pixels. Thus "SigmaSimpleColor" can be used almost everywhere.

We shall present below a faster version of the Sigma filter. Unfortunately, it is impossible to apply in this case the method used in the fast-averaging filter described in [1, p. 8] since the procedure is non-linear. The procedure can be made faster due to the use of a local histogram. The histogram is an array, in which each element contains the number of occurrences of the corresponding gray value or color intensity in the window. The sigma filter calculates the histogram for each location of the window by means of the updating procedure: gray values or color intensities in the vertical column at the right border of the window are used to increase the corresponding values of the histogram, while the values at the left border are used to decrease them:

Let OPP be the number of operations per pixel. The value $2 * 3 * W$ is the number of operations necessary to actualize the histogram and $2 * 3 * (2 * \text{tol} + 1)$ is the number of operations necessary to calculate the sum of $3 * (2 * \text{tol} + 1)$ values of the histogram and the corresponding number of used pixels. Thus, the overall number of operations is OPP $= 2 * 3 * W + 2 * 3 * (2 * \text{tol} + 1)$.

Here is the algorithm for the sigma filter with a local histogram for color images. As in the previous algorithm, "Image" and "Output" are three-dimensional fields.

Algorithm SigmaColor(Image[Width, Height, 3], hWind, Tol)

Step 1: Define the variables gv, xx, x1, X, y1, Y, yEnd, and yStart.
Step 2: Define the fields gvMin[3], gvMax[3], nPixel[3], and Sum[3].
Step 3: Define the two-dimensional field Histo[256, 3] and initialize it with 0.
Step 3: Set Y ← 0.
        Set Y ← Y + 1.
Step 4: Repeat until Y is less than Height:
    4.1: Set three variables: yStart ← Max(0, Y – hWind, 0),
                yEnd ← Min(Heigh – 1, Y + hWind), and X ← 0.
    4.2: Repeat until X is less than Width:
    4.2.2: Set c ← 0.
    4.2.3: Repeat until c is less than 3:
      4.2.3.1: If X is equal to 0, then:
    Begin 4.2.3.1
     Set gv ← 0.
     Repeat until gv is less than 256:
      Begin
       Set Histo[gv, c] ← 0.
       Set gv ← gv + 1.
      End   (End of Repeat until gv is less than 256.)
     Set y1 ← yStart.
     Repeat until y1 is less than yEnd:
      Begin y1
      Set xx ← 0.
      Repeat until xx is less than hWind:
       Begin
        Set Histo[Image[X,Y,c] ← Histo[Image[X,Y,c] + 1.
        Set xx ← xx + 1.
       End   (End of Repeat until xx is less than hWind.)
      Set y1 ← y1 + 1.
      End y1   (End of Repeat until y1 is less than yEnd.)
    End 4.2.3.1   (End If X is equal to 0.)
      4.2.3.2: If X is greater than 0, then:
    Begin 4.2.3.2
     Set x1 ← X + hWind.
     Set x2 ← X – hWind - 1.
     If x1 is less than Width – 1, then
      Begin x1
       Set y1 ← yStart.
      Repeat until y1 is less or equal to yEnd:
       Begin
        Set Histo[Image[x1,y1,c] ← Histo[Image[x1,y1,c] + 1.
        Set y1 ← y1 + 1.
       End   (End of Repeat until y1 is less or equal to yEnd.)
      End x1   (End of If x1 is less than Width – 1.)
     If x2 is greater or equal to 0, then
      Begin x2
       Set y1 ← yStart;
      Repeat until y1 is less or equal to yEnd:

```
              Begin
                  Set Histo[Image[x1,y1,c] ← Histo[Image[x1,y1,c] - 1.
                  If Histo[Image[x1,y1,c] is less than o, then return -1.
                  Set y1 ← y1 + 1.
              End         (End of Repeat until y1 is less or equal to yEnd.)
          End x2          (End of If x2 is greater or equal to 0.)
      End 4.2.3.2   (End of If X is greater than 0)
  4.2.3.3: Set Sum[c] ← 0.
  4.2.3.4: Set gvMin[c] ← Max(0, Image[X,Y,c] – Tol).
  4.2.3.5: Set gvMax[c] ← Min(255, Image[X,Y,c] + Tol).
  4.2.3.6: Set gv ← gvMin[c].
  4.2.3.7: Repeat until gv is less or equal to gvMax[c]:
              Begin
                  Set Sum[c] ← Sum[c] + gv*Histo[gv, c].
                  Set nPixel[c] ← nPixel[c] + Histo[gv, c].
                  Set gv ← gv + 1.
              End         (End of Repeat until gv is less …4.2.3.7)
  4.2.3.8: If nPixel[c] is greater than 0, then
                          set Output[X,Y,c] ← Sum[c]/nPoint[c].
              else set Output[X,Y,c] ← Image[X,Y,c].
  4.2.3.9: Set c ← c + 1.              (End of Repeat until c is less than 3, Step 4.2.3.)
  4.2.4: Set X ← X + 1.              (End of Repeat until X is less than Width, Step 4.2.)
  4.3: Set Y ← Y + 1.       (End of Repeat until Y is less than Height. Step 4.)
  Stop
```

The method "SigmaSimpleColor" is faster than "SigmaColor" if the width of the gliding window is less than 7 (hWind less than 3). At greater values of the width "SigmaColor" is faster. The working time of "SigmaColor" changes rather slowly with the width of the gliding window. Look please at Table 6.1.

The method "SigmaSimpleColor" and similar methods for grayscale images are used in almost all projects by the author.

6.1.2   Extreme Value Filter

The Sigma filter does not blur the edges. However, even an ideal edge can become somewhat blurred during the digitization of the image. The reason for this is that the

Table 6.1 Working times for a color image of 1200 × 1600 pixels in seconds with the Intel 50 GHz processor

	hWind			
Method	1	2	3	6
SigmaSimpleColor	0.71	1.68	3.11	10.26
SigmaColor	2.43	2.55	2.63	3.02

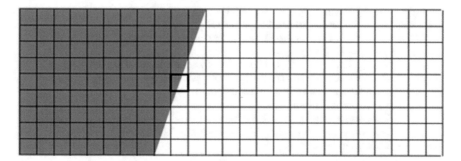

Fig. 6.5 The CCD matrix illuminated by a dark and a light area

boundary between a light and a dark area in an image projected to the set of light-sensitive elements, e.g., CCD, can fall occasionally near to the center of an element (Fig. 6.5).

Half of this element gets the brighter light, with the other half getting the darker one. Thus, this element obtains a light amount, which we call "middle value", laying between the values which are proper for the light and dark areas. The corresponding pixel has a middle value. It lies in the middle of the edge and the edge can be interrupted.

We suggest using the Extreme Value Filter to make the edges sharp. This filter applied to a gray level image finds two pixels in a small gliding window with the maximum and the minimum gray values. Then it calculates two differences between the gray value of the central pixel of the window and the maximum and minimum values and decides which of these two values is closer to the central value. The closer value is assigned to the central pixel in the output image. The edges become sharp. Here is the algorithm of the extreme filter for gray-scale images.

Algorithm ExtremGray(image Input[width, height], hWind)

Step 1: Declare the variables gv, gmax, gmin, xEnd, xStart, yEnd, yStart,
                 X, x, Y, and y.
Step 2: Set Y ← 0.
Step 3: Repeat until Y is less than Image.height:
    3.1: Set yStart ← Max(0, Y – hWind) and yEnd ← Min(Image.height, Y + hWind).
    3.2: Set X ← 0.
    3.3: Repeat until X is less than Image.width:
    3.3.1: Set five variables: yStart ← Max(y - hWind, 0), gmin ← 256,
        yEnd ← Min(y + hWind, height - 1), gmax ← 0 and y ← yStart.
    3.3.2: Repeat until y is less or equal to yEnd:
    3.3.2.1: Set x ← xStart.
    3.3.2.2: Repeat until x is less or equal to xEnd:
    Begin
    Set gv ← Input[x, y].
    If gv is less than gmin, then set gmin ← gv.
    If gv is greater than gmax, then set gmax ← gv.
    Set x ← x + 1.
    End   (End of repetition 3.3.2.2).
    3.3.2.3: Set y ← y + 1.   (End of repetition 3.3.2).
    3.3.3: If Input[X, Y] – gmin is less than gmax - Input[X, Y then
            set Output{X, Y] ← gmin.
       else set Output{X, Y] ← gmax.
    3.3.4: Set X ← X + 1.   (End of repetition 3.3).
    3.4: Set Y ← Y + 1.   (End of repetition 3).
Stop

We present below is the algorithm of the universal extreme method for color and grayscale images. The argument "nByte" must have the value 1 for grayscale images or 3 for color images. In this algorithm the image must be defined as a one-dimensional field of the size equal to nByte*width*height corresponding to the size of a grayscale image if nByte is equal to one, or corresponding to the size of a color image if nByte is equal to 3. The variables "width" and "height" should be global. The color channel C of the pixel with coordinates X and Y can be extracted from the field "Image" as

$$\text{value} = \text{Image}[C + nByte * (X + \text{width} * Y)].$$

If "nByte" is equal to one, then the repetition with C changing from 0 to nByte remains at a single value of C equal to 0 and the expression for value looks like.

$$\text{value} = \text{Image}[0 + 1 * (X + \text{width} * Y)] = \text{Image}[X + \text{width} * Y].$$

In this case "value" corresponds to the single value of the pixel (X, Y) of a gray level image.

The algorithm "ExtremUni" uses the utility algorithm "MaxC" calculating the lightness of the color (R, G, B) as the maximum of three values: G, 0.713*R, and 0.527*B.

Algorithm ExtremUni(Image[size], int hWind, int nByte)

Step 1: Define the variables C, CenterLight, K, MaxLight, MinLight, X and Y.
Step 2: Define the fields CenterColor[3], Color[3], Color1[3], Color2[3].
Step 3: Set Y ← 0.
Step 4: Repeat until Y is less than height:
   4.1: Set X ← 0.
   4.2: Repeat until X is less than width:
   4.2.1: Set C ← 0.
   4.2.2: Repeat until C is less than nByte:
      Begin
       Set four variables: Color2[C] ← Color1[C] ← Color[C] ←
           CenterColor2[C] ←Image[C+nByte*(X + width*Y].
       Set C = C + 1.
      End
   4.2.3: Set MinLight ← 1000 and MaxLight ← 0.
   4.2.4: Set K ← -hWind.
   4.2.5: Repeat until K is less or equal to hWind:
    4.2.5.1: If Y + K is greater or equal to 0 and Y + K is less than height, then
     4.2.5.1.1: Set J ← -hWind.
     4.2.5.1.2: Repeat until J is less or equal to hWind:
      4.2.5.1.2.1: If X+J is greater or equal to 0 and X+J is less than width, then
           Begin X+J
      4.2.5.1.2.1.1: Set C ← 0.
      4.2.5.1.2.1.2: Repeat until C is less than nByte:
           Begin
           Set Color[C] ← Image[C+nByte*(X+J +width*(Y+K))].
           Set C ← C + 1.
           End   (End of Repeat until C is less than nByte)

      4.2.5.1.2.1.3: If nByte is equal to 1, then set Light ← Color[0].
           else
           Set Light ← MaxC(Color[2], Color[1], Color[0]).
      4.2.5.1.2.1.4: If Light is less than MinLight, then
           set MinLight ← Light and Color1 ← Color.

      4.2.5.1.2.1.5: If Light is greater than MaxLight, then
           set MaxLight ← Light and Color2 ← Color.
           End X+J   (End of If X + J is greater …)

    4.2.5.1.2.2: Set J ← J + 1.   (End of Repeat until J is less …)
   4.2.5.2: Set K ← K + 1.   (End of 4.2.5, Repeat until K is less …)
   4.2.6: If nByte is equal to 1, then set CenterLight ← CenterColor[0].
    else Set CenterLight ← MaxC(CenterColor[2],CenterColor[1],CenterColor[0]).
   4.2.7: Set C ← 0.
   4.2.8: Repeat until C is less than nByte:
      Begin
      If CenterLight – MinLight is less than MaxLight – CenterLight, then
               Output[C + nByte*(X+width*Y] ← Color1[C].
      else   Output[C + nByte*(X+width*Y] ← Color2[C].
      Set C ← C + 1.
     End   (End of step 4.2.8, Repeat until C …)
   4.2.9: Set X ← X + 1. (End of repetition 4.2)
  4.3: Set Y ← Y + 1. (End of repetition Step 4)
Stop

6.2   The New Method of Edge Detection

We have published [1] a new efficient method of edge detection. Recently we have developed a simpler and faster version of these method.

We define edges as points where the gradient magnitude assumes a local maximum in the gradient direction. More precisely, this means that the edge near the point (X, Y) of a gray level image goes through the point (x, y) if the absolute value of the gradient has at (x, y) its maximum on the straight line going through (X, Y) parallel to the gradient.

However, in the case of color images there are *three gradients*: Gradient of the color channel Red, that of the channel Green, and that of the channel Blue. It is impossible to combine these three gradients to one since they may have quite different directions and magnitudes.

We suggest using the sum of the absolute differences of the three intensities of the channels: $SAD = |Green[x, y] - Green[x-1, y]| + |Red[x, y] - Red[x-1, y]| + |Blue[x, y] - Blue[x-1, y]|$. Instead of looking for the constrained maximum of the gradient we look for the maximum of SAD in each row during the scanning of the color image row by row. Later we repeat a similar procedure for the maximum of SAD in each column during the scanning of the color image column by column. The algorithm is described below. Before starting the edge detection, the image must be processes by the Sigma and the Extreme-value algorithms (Sects. 6.1.1 and 6.1.2). Then an image with doubled width and height must be defined representing a cell complex. This complex will serve as the output image for the algorithm of edge detection. The algorithm must have access to this complex.

The algorithm for edge detection obtains the image resulting from the Extreme-value algorithm as the input image. The algorithm scans the complex two times: first for all vertical cracks in all lines and then for all horizontal cracks in all columns. Consider first the scanning of the vertical cracks.

The repetition for vertical cracks tests each pair of pixels P_1 and P_2 horizontally adjacent to the crack and takes the colors of the corresponding pixels of the input image. The coordinates of the corresponding pixels are the coordinates of the pixels of the complex divided by two. Three differences of the intensities of the color channels dR, dG, and dB of the adjacent input pixels are calculated, and the absolute values of the differences are added. The sum is saved in the first element SAD_0 of the field SAD: $SAD_0 = |dR| + |dG| + |dB|$.

This field will contain the values of the absolute differences for fife subsequent pairs of pixels. They correspond to fife subsequent vertical cracks C_0 to C_4 (Fig. 6.6).

The fife values of SAD are compared with the given threshold. The values of SAD which are greater than the threshold are indications that the corresponding cracks *can* belong to the edge. The structure of the cracks whose SAD values are greater that the threshold is encoded in the bits of the integer variable Code: the bit $Code_i$ ($i = 0$ to 4) is set to 1 if the corresponding value of SAD_i is greater than the threshold.

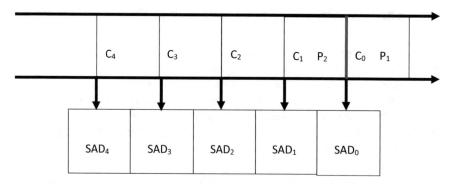

Fig. 6.6 Subsequent cracks and saved differences

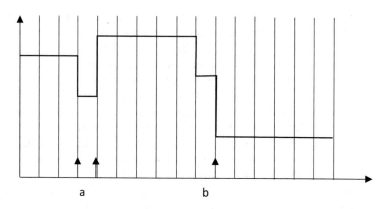

Fig. 6.7 Color differences with different signs

The algorithm for edge detection would be simpler if it would consider only the absolute differences of the color channels. However, then it could not distinguish between the two situations shown in Fig. 6.7.

For the edges to be thin, one should select among the adjacent cells the one with the maximum difference as the edge cell. But if two adjacent cells have differences with different signs, then both cells should be selected as edge cells. Therefore, it is necessary to consider the signs of the differences. The algorithm calculates the sum SAD of the absolute differences of the intensities of the color channels. The value SAD is always non-negative, i.e., it is either positive or zero. It is possible to assign a sign to the value of the difference: The sign should be equal to the sign of the difference of the intensities of the green channels of the pixels P1 und P2 if the latter difference is not equal to zero. Otherwise, it should be equal to the sign of the difference of the intensities of the red channels of the pixels P1 und P2 if the latter difference is not equal to zero. If, however, both differences green and red are equal to zero, then the sign should be equal to the sign of the difference of the intensities of the blue channels of the pixels P1 und P2. The latter difference is not equal to zero if the green and the red differences are zero, but the sum of three absolute differences is greater than the threshold. The sign is calculated by the function SignDif(Color1, Color2) where Color1 and Color2 are the colors of the pixels P1 and P2. The signed color difference DIF is then the product of SAD with the sign returned by the function SignDif(Color1, Color2). The signed differences are saved in the field DIF[5].

To decide whether a crack with the coordinate X belongs to the edge, the algorithm must test two cracks before X and two after X as explained above. For each of these cracks both the absolute difference of the colors SAD and the signed difference DIF of two pixels adjacent to the crack must be calculated and saved in the fields SAD[5] and DIF[5] correspondingly.

The easiest way would be to do a repetition loop for the value of the coordinate Crack.X changing from 2 until width—3 (the values of Crack.X for which both adjacent pixels lie in the complex) and to calculate in another repetition with the variable I from 0 to 4 the color differences for the two pixels adjacent to the crack with $X1 = Crack.X + 4 - 2* I$. But then each difference would be computed five times for each value of X.

In order not to calculate the color difference near a crack multiple times, the algorithm makes the repetition for changing the variable Crack.X from the value of -2 to width $- 1$ thus overlapping all five locations adjacent to each of the tested cracks. Both the absolute and the signed differences of the adjacent colors for $X = Crack.X + 4$ are saved in the elements SAD[0] and DIF[0] of the corresponding fields and the content of both fields SAD and DIF are shifted by one element to higher indexes at the end of the repeated processing. The differences are only calculated if both pixels with the coordinates $X - 1$ and $X + 1$ are in the complex. Otherwise, both differences are set equal to 0. At the beginning of the line, all elements of both fields SAD and DIF field are set to 0.

The decision as to whether the crack C_i belongs to the edge is made with the aid of the look-up table LUT. LUT is an integer field of 32 elements corresponding to the 32 possible values of the five bit variable Code. The values of LUT lie between 0 and 8. They denote the 9 possible situations corresponding to the values of the variable Code. For example, LUT[Code] equal to 0 means that the crack C_2 does not belong to the edge. The value LUT[Code] equal to 1 correspond to the values of Code equal to 12 or 13. The decimal value 12 corresponds to the binary value Code (binary) $= (01100)$ which means that SAD_3 and SAD_2 are greater than the threshold. If the additional condition that SAD_2 is greater than SAD_3 is fulfilled, then the crack C_2 belongs to the edge. These conditions can be seen in the description of the logical function ToEdge(lut, DIF) below.

Similar calculations and decisions are performed in the second repetitions for horizontal cracks and can be seen in the following description of the algorithm EdgeDetect. The algorithm obtains as parameters the input image "Input", the integer value "Thresh" which is the threshold for the values of the field SAD, and the integer value "nByte" which is equal to 1 for gray-scale images and equal to 3 for color images. Thus, the algorithm can be used both for grayscale and for color images. The parameter "nByte" makes the change between grayscale and color as described above in Sect. 6.1.2 before the description of the algorithm "ExtremUni".

To make this change possible, the image must be defined as a one-dimensional field of the size nByte*width*height, and the access to the color channel C, C = 0, 1, 2; of the pixel (X, Y) will be made as "value = Complex[C + nByte*(X + width*Y)]". The values "width" and "height" are global. The output of the algorithm is a two-dimensional complex of the size (2*width + 1)*(2*height + 1).

The algorithm has a direct access to the output image "Output".

Algorithm EdgeDetect(Input, Thresh, nByte)

Step 1: Define the variables C, Code, I, NX, X, X1, Y and Y1.
Step 2: Define the fields DIF[5], SAD[5], Color1[3], Color2[3].
Step 3: Define and initialize the field Bit ← {0, 1, 2, 4, 8, 16}.
Step 4: Define and initialize the field LUT ← {0, 0, 0, 0, 2, 2, 3, 5, 0, 0, 0, 0,
    1, 1, 4, 9, 0, 0, 0, 0, 2, 2, 3, 5, 0, 0, 0, 0, 6, 6, 7, 8}.
Step 5: Set Y ← 1.
Step 6: Repeat until Y is less than height:
    6.1: Set all elements of DIF and SAD equal to 0.
    6.2: Set X ← -2.
    6.3: Repeat until X is less than width - 1:   (X is abscissa of a crack)
    Begin
    Set X1 ← X + 4.
    If X1 + 1 is less than width, then:
      Begin
        Set SAD[0] ← 0 and C ← 0.
        Repeat until C is less than nByte:
          Begin
            Set Color1[C]←Input[C + nByte*((X1+1)/2 +width*(Y/2)]. (Standard coord.)
            Set Color2[C]←Input[C + nByte*((X1-1)/2 +width*(Y/2)].
            Set SAD[0]←|Color1[C] – Color2[C]|;   (Absolute difference).
            Set C ← C + 1.   (End of Repeat until C ...)
          End
        Set DIF[0] ← SAD[0]*SignDIF(Color1, Color2);
      End
    else Set DIF[0] ← SAD[0] ← 0. (End "If X1 + 1 is less than width)
    Set I ← 0.
    Repeat until I is less than 5:
      Begin
        If (SAD[I] is greater than Thresh set the Ith bit of Code.
        Set I ← I + 1.
      End
    If the function ToEdge(LUT[Code], DIF) returns true, then
      Begin
        Set Output[X, Y] ← 1.   (The vertical edge crack)
        Set Output[X, Y-1] ← Output[X1, Y-1] + 1. (its upper point)
        Set Output[X, Y+1] ← Output[X1, Y+1] + 1   (its lower point)
      End
    Set I ← 4.
    Repeat as long as I remains greater than 0:

Begin
      SAD[I] ← SAD[I – 1].
      DIF[I] ← DIF[I – 1].
      Set I ← I – 1.
   End   (End Repeat as long as I remains greater than 0.)
  Set X ← X + 2.   (Next crack)
  End   (End repetition unless X less than width – 1. 6.3)
  6.4: Set Y ← Y + 2.   (Next row of pixels and end of repetition 6)
Step 7: Set X ← 1.   (Start of the repetitions for horizontal cracks)
Step 8: Repeat until X is less than width:
  8.1: Set Y ← -2.
  8.2: Repeat until Y is less than height - 1:
 Begin 8.2
  Set Y1 ← Y + 4.
  If Y1 + 1 is less than height, then:
   Begin
    Set SAD[0] ← 0 and C ← 0.
    Repeat until C is less than nByte:
     Begin
      Set Color1[C]←Input[C+nByte*(X/2+width*(Y + 1)/2)].   (Standard coord.)
      Set Color2[C]←Input[C+nByte*(X/2+width*(Y - 1)/2)].
      Set SAD[0]←|Color1[C] - Color1[C]|;   (Absolute difference).
      Set C ← C + 1. (End of Repeat until C)
     End
    Set DIF[0] ← SAD[0]*SignDIF(Color1, Color2).
   End
  else Set DIF[0] ← SAD[0] ← 0. (End "If Y1 + 1 is less than height".)
  Set I ← 0.
  Repeat until I is less than 5:
   Begin
    If (SAD[I] is greater than Thresh set the Ith bit of Code.
    Set I ← I + 1.
   End   (End Repeat until I is less than 5.)
  If function ToEdge(LUT[Code], DIF) returns true, then
   Begin
    Set Output[X, Y1] ← 1.   (The vertical edge crack)
    Set Output[X - 1, Y1] ← Output[X - 1, Y1] + 1. (its left point)
    Set Output[X + 1, Y1] ← Output[X + 1, Y1] + 1. (its right point)
   End
  Set I ← 4.
  Repeat as long as I remains greater than 0:
   Begin
    SAD[I] ← SAD[I – 1].
    DIF[I] ← DIF[I – 1].
    Set I ← I – 1.
   End
  Set Y ← Y + 2.   (End of repetition 8.2)
 End 8.2
  8.3: Set X ← X + 2.   (End of repetition 8)
Stop

The algorithm EdgeDetect uses the integer function "SignDif" and the logical function "ToEdge". Here are the corresponding algorithms.

Algorithm SignDif(Color1, Color2)

Step 1: Define the variables SignDif, iColor1 and iColor2.
Step 2: Set iColor1 ← Color1.Green*256^2 + Color1.Red*256 + Color1.Blue.
Step 3: Set iColor2 ← Color2.Green*256^2 + Color2.Red*256 + Color2.Blue.
Step 4: If iColor1 – iColor2 is negative, then set SignDif ← -1;
      else set SignDif ← 1.
Stop

Algorithm ToEdge(integer Lut, Field DIF)
Step 1: Define and initialize the logical variable ToEdge ← false.
Step 2: If Lut is equal to 2 OR Lut is equal to 3 AND
      DIF[2]*DIF[1] is positive AND |DIF[2]| is greater or equal to |DIF[1]| OR
  Lut is equal to 1 AND DIF[2]*DIF[3] is positive AND
     |DIF[2]| is greater than |DIF[3]| OR DIF[2]*DIF[3] is negative OR
  Lut is equal to 4 AND DIF[2]*DIF[3] is positive AND
     DIF[2]*DIF[1] is positive AND |DIF[2]| is greater than |DIF[3]| AND
     |DIF[2]| is greater or equal to |DIF[1]| OR DIF[2]*DIF[1] is negative AND
     DIF[2] * DIF[3] is negative OR DIF[2]*DIF[1] is positive AND
     SAD[2] is greater than or equal to SAD[1] AND DIF[2] * DIF[3] is negative OR
  Lut is equal to 5 AND DIF[2] * DIF[1] is positive AND
     SAD[2] is greater than or equal to SAD[1] AND
     DIF[2] * DIF[0] is positive AND SAD[2] is greater than or equal to SAD[0] OR
     DIF[2] * DIF[1] is negative AND DIF[2] * DIF[0] is positive AND
     SAD[2] is greater than or equal to SAD[0] OR
     DIF[2] * DIF[1] is negative AND DIF[2] * DIF[0] is negative OR
     DIF[2] * DIF[1] is positive AND SAD[2] is greater than or equal to SAD[1] AND
     DIF[2] * DIF[0] is positive OR
  Lut is equal to 6 AND DIF[2] * DIF[3] is positive AND
     SAD[2] is greater than SAD[3] AND DIF[2] * DIF[4] is positive OR
     SAD[2] is greater than SAD[4] OR
     DIF[2] * DIF[3] is negative OR DIF[2] * DIF[3] is positive AND
     SAD[2] is greater than SAD[3] AND DIF[2] * DIF[4] is negative OR
  Lut is equal to 7 AND NOT (DIF[2] * DIF[1] is greater than or equal to 0 AND
     SAD[2] is less than SAD[1] OR
     DIF[2] * DIF[3] is positive AND SAD[2] is less than or equal to SAD[3] OR
     DIF[2] * DIF[3] is positive AND DIF[2] * DIF[4] is positive AND
     SAD[2] is less than or equal to SAD[4]) OR
  Lut is equal to 8 AND ((DIF[2] * DIF[1] is negative OR
     SAD[2] is greater than or equal to SAD[1] AND
     (DIF[2] * DIF[0] is less than 0 OR SAD[2] is greater than or equal to SAD[0])) AND
     (DIF[2] * DIF[3] is negative OR SAD[2] is greater than SAD[3] AND
     (DIF[2] * DIF[4] is negative OR SAD[2] is greater than SAD[4]))) OR
  Lut is equal to 9 AND NOT(DIF[2] * DIF[1] is positive AND
     DIF[2] * DIF[0] is positive AND SAD[2] < SAD[0] OR
     DIF[2] * DIF[1] is positive 0 AND SAD[2] is less than SAD[1] OR
     DIF[2] * DIF[3] is positive 0 AND SAD[2] is less than or equal to SAD[3]),
  then set ToEdge ← true.
Stop

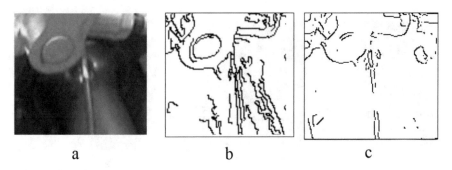

a   b   c

Fig. 6.8 Fragment of the image (**a**), edges detected by our method (**b**) and edges detected by the Canny edge detector (**c**)

To demonstrate the success of our method of edge detection we have copied the images from the article "Canny edge detector" in en.Wikipedia and applied our method to the color image from that article.

Figure 6.8 shows a fragment of this image (a), the edges detected by our method (b) and the edges detected by the Canny edge detector [6].

The Canny algorithm for edge detection demands that a color image should be converted to a grayscale image. However, after this conversion some edges in the color image disappear. This occurs if two adjacent areas have different colors but the same lightness.

Our algorithm can detect true color edges between two areas with the same lightness but different colors. For this purpose, one needs the Sigma filter and the Extreme filter suitable for color images. Sigma filter for color images is described above in Sect. 6.1.1; the Extreme filter for color images - in Sect. 6.1.2.

Our edge detector calculates, as described above, the color difference between two adjacent pixels $P_1(Red_1, Green_1, Blue_1)$ and $P_2(Red_2, Green_2, Blue_2)$ as the sum of the absolute differences of the color channels. The edge lies between two pixels if their absolute color difference is greater than a predefined threshold. The advantage of our method can be seen on edges between two areas with different colors and equal lightness's as for example in the following image Fig. 6.9 between the red flower and a green leaf having almost the same lightness.

Please pay attention to locations indicated by the arrows. As it can be seen, many edges are missing in Figure 6.9d.

6.3   Applications of Edge Detection

We present here three examples of the applications of detecting the edges:

1. Image compression.
2. Recognition of circular objects.
3. Recognition of bicycles in street images.

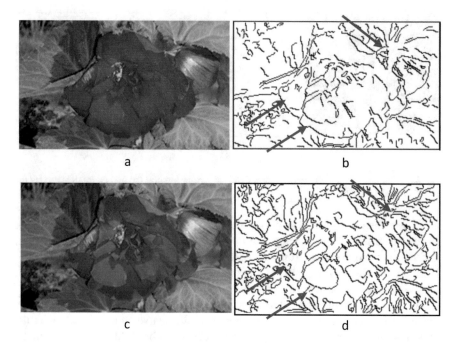

Fig. 6.9 Original image (**a**); edges of the color image (**b**); original converted to gray values (**c**); edges of the gray-scale image (**d**)

6.3.1   Image Compression by Means of Edge Detection

A new method of image compression based on edge detection is described in [1]. The idea is based on the supposition that important information in an image is in pixels lying near the edges. All other pixels contain only a homogeneous distribution of colors which can by reconstructed by solving numerically a partial differential equation with the methods of finite differences. We use here edge detection with the method described in the previous Sect. 6.2. We represent edges as an abstract cell complex (ACC) described in Chap. 2. This representation has advantages because an element of the edge is defined by the *difference of colors of two adjacent pixels*. Thus, an element of the edge belongs to neither of these two pixels but rather to a space element lying between these pixels. Such a space element does not exist in a digital image; however, it exists in an abstract cell complex. This is the one-dimensional cell called "crack", while pixels are two-dimensional cells. The ends of a crack are zero-dimensional cells called points. When considering pixels as small squares, then cracks are the sides of these squares; points are their corners and end points of the cracks (Fig. 6.10).

The edges compose a network of thin lines being sequences of cracks and points where each line segment is either a closed curve or a sequence starting and ending

Fig. 6.10 Example of edges

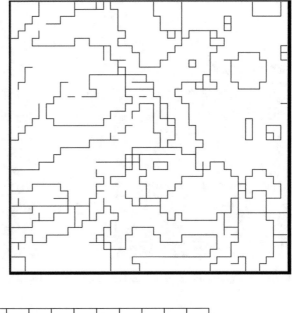

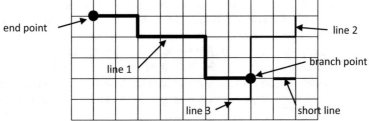

Fig. 6.11 Examples of a line (bold segments), a branch point and a short line

either at an end point or at a branching point where three or four line segments meet (Fig. 6.11 above).

To completely describe the edges of an image one needs only a list of lines. To reconstruct an image, it is sufficient to know the colors of pixels incident to the cracks of each line. However, these colors are almost constant along one side of a line. If there were great changes of colors in the set of pixels lying at one side of a line, then there were also edge cracks perpendicular to the line but there are no such cracks. Because the colors at one side of a line changes slowly, it is sufficient to save for each side of a line colors in one pixel at the beginning and one at the end of a line. All other colors at one side of a line can be calculated by means of an interpolation between these colors at the beginning and at the end of a line.

To perform this interpolation, one needs to know the exact location of all elements of the line. All these locations can be specified by the coordinates of the starting point and by the sequence of the directions of all cracks of the line. A crack of an oriented line (this means, it is specified which point is the starting one) can have

Fig. 6.12 Example of an original and a reconstructed image. Compression rate 51

one of four directions: direction 0 goes to the right, direction 1 goes downwards, direction 2 goes to the left and direction 3 goes upwards. In this way the edge lines of an image and the adjacent colors can be encoded very sparingly. Also, colors at the borders of the image are necessary for the reconstruction. To know this colors, it is enough to know only the colors at the four edges of a rectangular image. All other colors at the borders can be calculated by an interpolation. Also, colors at the cracks of the edge crossing the sequence of border pixels should be regarded during this interpolation.

The list of the edge lines with the adjacent colors is the compressed representation of the image. Its volume is essentially smaller than the volume of the original image. Nevertheless, the image can be reconstructed from the list of the edge lines. The rate of the compression is the division of the volume of the original image divided by the volume of the list of the edge lines. It is sometimes greater that the compression rate of the widely used compression by means of the JPG method. The image can be reconstructed from the list of the edge lines.

An example of a reconstructed image is shown in Fig. 6.12.

As the reader can see, the quality of the reconstruction is good. The compression rate for this image is 51. The JPEG compression rate (at 75% quality) is 43.

6.3.2   *Recognition of Circular Objects*

Recognition of circular objects can be made by representing curves in two-dimensional digital images as polygons as described in Sect. 5.4 for boundaries.

a   b

Fig. 6.13 Example of an image (**a**) and the results (**b**)

The same method can be used for edge curves. This kind of curve representation is useful for image analysis because the shape of a polygon can be easily investigated by simple geometrical means such as measuring lengths and angles. Polygonal approximation also suggests a new method of estimating curvature of digital curves. For this purpose, a polygon can be replaced by a smooth sequence of circular arcs and straight-line segments. "Smooth" means that each straight segment is the tangent to the previous and to subsequent arc.

For the recognition of objects whose shape of the edges is like a circle we use a modified method of least squares described in Sect. 5.4.4. The method was developed by the author some years ago for checking the quality of solder bumps in wafers. The method calculates the estimates of the radius and of the coordinates of the center. Figure 6.13 shows an example of using the method for sorting apples according to their sizes.

Figure 6.14 shows an example of recognized circles of wafer bumps. Recognized circles are drawn as red lines.

It is important to draw the attention of the reader to the possibility of our method to recognize circles even if they are presented only by disconnected pieces of arcs as shown in Fig. 6.15 showing a fragment from Fig. 6.14.

The method calculates the center and the radius of a the optimally suitable circle for each piece and joins pieces that have similar centers and radii. Then the optimal circle for the set of joined pieces is calculated.

6.3.3   Recognition of Bicycles in Street Images

Due to the good properties of our method for circle recognition we came to the idea to use this method to recognize the wheels of bicycles which are ideal circles. However, if the bike is positioned so that the plane of its frame makes an acute angle with the viewing direction, then the wheels look like ellipses, rather than like circles. Therefore, we also need a method of recognizing ellipses. Unfortunately, we have not succeeded with generalizing our method of circle recognition mentioned in

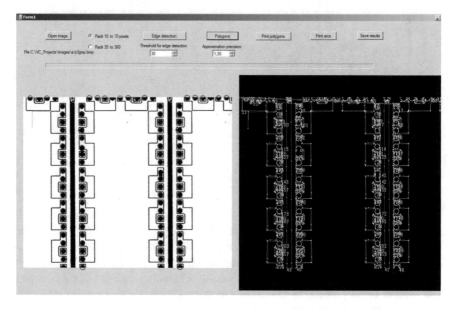

Fig. 6.14 Example of recognized 93 circles of wafer bumps

Fig. 6.15 Disconnected pieces of arcs representing a circle

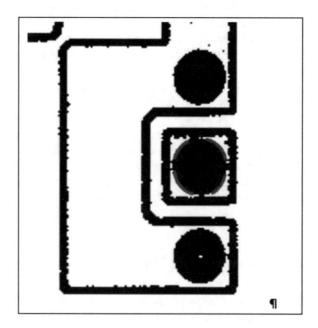

the previous section for ellipses. Since an ellipse is defined by only a small number of parameters, namely by five, it is possible to use the classical procedure of least squares for the recognition of ellipses represented by pieces of edge curves.

Fig. 6.16 Examples of images with recognized bicycles

We have described the method of recognizing bicycles in street images in [1], Chap. 13. The method approximates the edges with polygons, dissolves each polygon in arcs (as pieces of polygons with limited angles between adjacent polygon edges) and joins arcs having similar curvature and similar positions of curvature centers. Ellipses are recognized for groups of joined arcs. If their centers and sizes of ellipse axes correspond to certain conditions, then the ellipses are accepted as bicycle wheels. Then edges of polygons are tested for their locations and orientation relative to the locations and sizes of the wheels. Pieces of polygon edges satisfying certain conditions can be regarded as parts of the frame of a bicycle.

Figure 6.16 shows some examples of images with recognized bicycles.

References

1. Kovalevsky V. Modern algorithms for image processing. New York: Apress; 2018. ISBN 978-1-4842-4247-7
2. Lee JS. Digital image smoothing and the sigma filter. Comput Vision Graph Inform Process. 1983;24(2):255–69.

3. Chochia PA. Image enhancement using sliding histograms. Comput Vision Graph Inform Process. November 1988;44(2):211–29.
4. Tomasi C, Manduchi R. Bilateral filtering for gray and color images. In: Proceedings of the IEEE international conference on computer vision. Piscataway: IEEE Press; 1998. p. 839–46.
5. Dellaert F. The expectation maximization algorithm. CiteSeerX 10.1.1.9.9735; 2002.
6. Canny I. A computational approach to edge detection. IEEE Trans Pattern Anal Mach Intell. 1986;8(6):679–98.

Chapter 7
Surface Traversing and Encoding in 3D Images

Abstract This chapter describes two algorithms of traversing surfaces in 3D images and the theory of subdividing a surface into digital plane patches (DPP). The algorithm "SpiralTracing" is the only known algorithm which can encode a surface as a sequence of adjacent facets while other algorithms encode a surface as a not ordered set of facets. So does the here described algorithm "CORB_3D". This chapter also describes the theory of digital plane patches.

Keywords Surface · Traversing surfaces · Spiral tracing · Genus of a surface · Reversible tracing · Efficiency of encoding · Algorithm CORB_3D · Digital plane patches (DPP) · Separation form · Bases of a DPP · Semi-singular and singular orientation of facets · Segmentation of a surface into DPPs · Recognition of a DPP · Choice of appended facets

A three-dimensional digital image is a set of voxels which we can consider as small cubes. Detecting boundaries of subsets in a three-dimensional image makes similar problems as in the case of two-dimensional images; the solution of the problems is also similar: It is necessary to construct and to consider a 3D cell complex parallel to the 3D image.

As in the two-dimensional case, the boundary of a three-dimensional subset is the set of cells whose smallest neighborhood intersects both the subset and its complement. Let us consider some details of a three-dimensional complex.

A three-dimensional, or 3D complex, contains cells of dimensions from zero to three. Three-dimensional cells are the voxels. In a Cartesian complex a voxel looks like a cube: it possesses six sides which are the two-dimensional cells called facets. They can also be called pixels. A facet possesses four sides which are one-dimensional cells or cracks. The end points of a crack are the zero-dimensional cells or points.

Smallest neighborhoods of cells are important for the recognition of boundaries. The smallest open neighborhood of a voxel is the voxel itself. This neighborhood consists of a single cell. Therefore, this neighborhood cannot intersect two

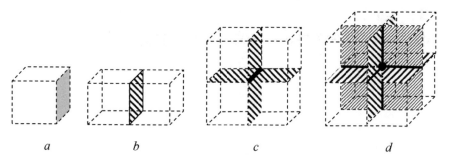

a   *b*   *c*   *d*

Fig. 7.1 Smallest neighborhoods of cells in a three-dimensional Cartesian complex; neighborhood of a voxel (**a**); of a facet (**b**); of a crack (**c**) and of a point (**d**)

disjoint sets: the subset and its complement. Thus, the boundary of a 3D set contains no voxels.

The smallest open neighborhood of a facet contains the facet itself and two voxels while the facet is the common side of these two voxels (Fig. 7.1b). One of these voxels can belong to the subset while the other one belongs to the complement. In such a case the facet belongs to the boundary.

The smallest open neighborhood of a crack (Fig. 7.1c) contains the crack itself, four voxels and four facets while the crack is the common side of these four facets. Thus, a crack can also belong to a boundary.

The smallest open neighborhood of a point contains the point itself, eight voxels, 12 facets and six cracks while the point is the common end point of these six cracks (Fig. 7.1d). A point can also belong to a boundary. Therefore, the boundary of a 3D subset consists of facets, cracks, and points. It contains no voxels.

There are several possibilities to trace and encode the surfaces of bodies in a tree-dimensional image. Thus Rosenfeld et al. [1] describe an algorithm for tracing and encoding a surface. The authors regard a 3D digital binary image as a set of upright unit cubes called voxels whose vertices have integer coordinates. They regard a surface as a set of facets, which are unit squares shared by a voxel of the foreground and a voxel of the background. Thus, they regard the surface as the frontier of a 3D subset of a 3D Cartesian complex (without calling it as such by name). Further, they introduce the notion of *strongly edge-adjacent facets*. The definition in their p. 306 is rather difficult to understand. To translate it into the language of complexes we consider first the notion of a regular surface.

The authors introduce the notion of a *regular surface* as a maximal set of pairs of adjacent facets having a common non-singular incident crack or incident to the same foreground voxel. They have proved that each facet of a regular surface has exactly four facets edge-adjacent to it and that for any 6-component L of the foreground V and any 18-component M of the background B such that their closures intersect at some facets, the frontier of V is a regular surface.

They briefly describe two algorithms for traversing the set of all facets of a regular surface. The algorithms are based on regarding a regular surface as a graph whose vertices are the facets of the surface and whose edges connect pairs of vertices corresponding to strongly edge-adjacent facets. One of the algorithms is the well-known breadth-first graph traversal algorithm. The second one is that suggested by Gordon and Udupa [2]. It is faster than the first one since the graph is replaced by a digraph and only approximately 1/3 of the facets are visited twice.

The authors of [1] also suggest representing a regular surface by the Euler circuit of the graph representing the surface. As known, a Euler circuit of a graph is a closed path going through all vertices of the graph that passes through each edge of the graph exactly once. It is well-know that a graph whose each vertex is incident to an even number of edges, possesses a Euler circuit. This representation is rather economical since, to trace the graph, it is only necessary to indicate the position of each edge of the graph relative to the position of the previous edge in the sequence rather than to indicate the coordinates of each vertex. The vertices of the graph representing a surface are the facets. Its edges are the cracks lying between two adjacent facets. The authors describe no algorithm for constructing the Euler circuit, but rather provide a reference to [3].

Thus, we see that the notion of a regular surface can be employed for efficiently tracing and encoding 3D binary images. The drawback of this approach is that the produced code correctly represents components of an image if they are strongly connected (6-connected). However, in practice we are often confronted with other images. Therefore, we have developed three methods for tracing and encoding surfaces in any images. They are described in Sect. 13 of [4]. There are among them: a simple algorithm using the well-known depth-first search in the adjacency graph of a T-component, an algorithm finding and economically encoding a Euler circuit for each T-component, a still more economical algorithm encoding a multivalued 3D image by means of tracing the frontiers of parallel slices of a 3D set, and an algorithm called "Spiral Tracing" tracing and encoding a surface and detecting its genus. For example, the genus of a sphere is zero, the genus of a torus is one, and the genus of a surface with n tunnels is n. All algorithms detect the components of the image. We describe below the algorithm "Spiral Tracing" which makes an economical coding of each component of the set of the boundaries.

7.1 Algorithm "Spiral Tracing"

(Section 7.1 was quoted from the author's book [4] with the permission of the publisher).

7.1.1 The Idea of the Spiral Tracing

This is an efficient method [4] producing a single connected sequence of code elements for each closed surface. Most of the facets of a surface of genus 0 is visited only once. According to the method an arbitrary facet of the surface B must be

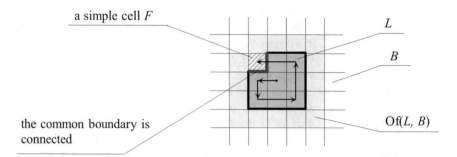

a simple cell *F*

L

B

the common boundary is
connected

Of(*L*, *B*)

Fig. 7.2 The moves at the beginning of the tracing

selected as the starting facet and its closure must be labeled. Then the opening
frontier Of(*L*, *B*) ([4], Definition OF, Section 3.1, p. 38) of the labeled set *L* of cells is
traced, its facets are encoded (1 byte per facet), and closures of simple facets are
labeled one after another. "Simple" means that the intersection $\partial F \cap \partial L$ of the facet's
boundary ∂F with the boundary ∂L of *L* is not empty and connected (it is a 1-ball),
while the complement $\partial F\text{-}\partial L$ is also not empty and connected. This ensures that
L remains homeomorphic to a closed 2-ball ([4], Definition BM, Section 3.4, p. 48).
If the surface *B* is homeomorphic to a sphere (genus $= 0$), then the traced sequence is
in most cases a Hamilton path in the adjacency graph of the facets: Each facet is
visited exactly once. In some surfaces of genus 0 a small portion of the facets is
visited twice (Fig. 7.2).

In surfaces of genus greater than 0 there always remain a few non-simple facets
which are visited twice. Their code elements are attached to the end of the sequence
of simple facets. Consequently, the code sequence is always connected and is thus
economical.

A brief verbal description of the algorithm follows at the end of this section. A
detailed description is presented in the next section.

During the tracing of the current frontier of *L* the relative positions of joined
facets are stored as a three-dimensional chain code as explained below. This
procedure must be repeated as often as there are unlabeled simple facets incident
to the cracks of the frontier of *L*. If the next facet encountered during the tracing is a
non-simple one, then it is not joined with *L*, i.e., it is not labeled. However, its
position is nevertheless included in the code to make the sequence of the encoded
facets continuous. At the beginning of the process the trajectory of the tracing
evolves like a spiral. The sequence of grasped facets looks like the peeling of a
potato.

The tracing stops when there are no more simple facets incident to the frontier of
L. At this stage, all facets of *B* are grasped in the code (some of them more than
once). This code can be employed as the code of the surface *B*. The code specifies the
surface *B* and therefore also the set *V* uniquely: The set *V* can be exactly
reconstructed from the code by one of the methods described in [4], Sect. 11.4.

Fig. 7.3 Encoding the
directions of the coordinate
axes and that of the cracks

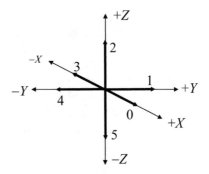

Theoretical background of these ideas is to be found in [4], pages from 281 to 283.

We denote the sequence of all codes obtained during *a full tracing* of the frontier $Fr(L, B)$ after the last simple facet was found as the rest sequence.

The algorithm:

1. Take any facet of B as the starting facet F_0, label its closure, and save its coordinates as the starting coordinates of the code. This is the seed of L. Denote any one crack of the frontier of $Fr(F_0, B)$ as C_{old} and find the facet F of B that is incident to C_{old} and adjacent with F_0. Set F_{old} equal to F_0 and the logical variable *REST* to FALSE. *REST* being equal to TRUE would indicate that the tracing of the rest sequence runs.
2. (Start of the main loop) Find the crack C_{new} as the first unlabeled crack of $Fr(F, B)$ encountered during the scanning of $Fr(F, B)$ clockwise, while starting with the end point of C_{old} which is in $Fr(L, B)$. If there is no such crack and F is labeled, then stop the Algorithm: The encoding of B has been completed.
3. If F is simple, then label its closure.
4. Put the code of the direction of the movement from F_{old} to C_{old} (3 bits) and that of the movement from C_{old} to F (next 3 bits) into the next byte of the code. Directions are shown in Fig. 7.3 above. If facet F is non simple, set the corresponding bit in the code (to recognize codes of non-simple facets in the ultimate sequence).
5. If *REST* is TRUE, then check whether F is equal to F_{stop} and simultaneously C_{new} is equal to C_{stop}. (These variables can be defined in item 6 during previous steps). If it is the case, then stop the Algorithm and analyze the rest sequence to specify the genus of B as explained below. Delete multiple occurrences of facets from the rest sequence.
6. If F is simple, then set *REST* equal to FALSE. If F is not simple, then set F_{stop} equal to F, C_{stop} equal to C_{new}, and *REST* equal to TRUE.
7. Set F_{old} equal to F. Find the facet F_{new} of B incident to C_{new} and adjacent to F. Set F equal to F_{new} and C_{old} equal to C_{new}. Go to item 2.

End of the algorithm

As seen from the Algorithm, after the last simple facet has been found, the tracing of Fr(L, B) is continued and the non-simple facets are recorded (in the rest sequence) until the path along the frontier Fr(L, B) becomes closed. The genus G of the surface is computed by means of the Euler number of the rest sequence. The Euler number E of a surface is the number of facets plus the number of points minus the number of cracks. The genus G of a surface is equal $G = (2 - E)/2$. The Euler number can be deduced from the numbers of non-simple facets which have been visited three or four times.

Theorem GEN The genus G of B can be deduced from the properties of the rest sequence. If the rest sequence is empty, then $G = 0$; otherwise, it is equal to:

$$G = (1 + N_3/2 + N_4)/2 \tag{7.1}$$

where N_3 is the number of facets which occur in the rest sequence three times and N_4 is the number of facets which occur in the rest sequence four times.

Proof of Theorem GEN To specify the genus G of B we first calculate its Euler number. The Euler number of L is equal to 1 since L was developed from a single facet by means of an operation which does not change the Euler number. If the rest sequence is empty, then B is the union of the closed 2-ball L with the last facet, which is an open 2-ball. Thus, in this case the Euler number is 2 and $G = 0$.

If the rest sequence $B-L$ is not empty, then B is a union of L and $B-L$. The set $B-L$ consists of facets incident with Fr(L, B) and cracks lying between these facets. It is a stripe one facet wide (compare Fig. 7.6 below). It can be regarded as a union of loops, like those shown in Fig. 7.6 below. The Euler number of a loop is zero since it consists of an equal number of facets and cracks. If two loops intersect (compare Fig. 7.6), then they have P facets and $P-1$ cracks in common (the cracks of the boundary Fr(L, B) do not belong to the loops). The count $P - (P - 1) = 1$ of these cells must be subtracted from the sum of the Euler numbers of the loop being zero. Thus, each location containing an intersection of two loops reduces the Euler number by 1. On the other hand, at each intersection there is either one facet with 4 adjacent facets in the loops or two facets with 3 adjacent facets. Thus, the number N_4 or half the number N_3, i.e., $N_3/2$, must be subtracted from the Euler number of the loops. Therefore, the Euler number E of B is equal to $E = 1 - N_3/2 - N_4$ and the genus G of B is equal to $G = (2 - E)/2 = (1 + N_3/2 + N_4)/2$.

The run time of the algorithm is obviously proportional to the length of the code, since the time necessary to produce a code element is constant. The length of the code can be essentially greater than the number of facets in B since the non-simple facets can arrive in the code many times. The worst case take place for sets of voxels looking like a dumbbell: Two cylinders of the same diameter connected at their flat ends by a long thin rod. After the facets in the surface of the rod are labeled except for a sequence one facet wide, the tracking can run along the rod back and forth as many times as the number of facets in the height of the cylinders. Thus, the worst-case complexity is $O(N^2)$, N being the number of facets in B.

There is a possibility to make the time complexity linear in N. This can be reached if the algorithm will change from the clockwise to the counterclockwise tracing mode each time when a non-simple facet occurs. This version of the algorithm is described below.

7.1.2   The Reversible Tracing

The only drawback of the method of the previous section is that the part of the frontier of L which is incident to non-simple facets must be sometimes traced repeatedly to reach all simple facets. This is due to the fact that in spite of the connectedness of $B-L$ the *subset of simple facets* in $B-L$ can be disconnected. Thus, the sequence can contain many non-labeled facets, and the sequence becomes unnecessarily long.

This difficulty can be overcome by means of *reversing the direction of the tracing* each time when a non-simple facet is encountered. This means that the tracing is being continued in such a direction that the labeled facets lie no more to the left-hand side of the direction (left tracing) but rather to the right-hand side (right tracing).

Figure 7.4a presents an example of the left tracing: The facets labeled before are shown as a shaded region. They lie to the left-hand side of the direction of the tracing. The facet „N"is a non-simple one: Its frontier crosses the frontiers of the facets which were labeled before at two disconnected locations, namely at facets 3 and 14.

Figure 7.4b demonstrates how the tracing has been continued as the right tracing was starting with facet 14 and running through facets 15–22.

When performing the reversible tracing, it is appropriate to trace not the usual frontier of the labeled region but rather the opening frontier ([4], Definition OF, Section 3.1, p. 38) of this region. Sequences of traced facets as shown in Fig. 7.4a, b

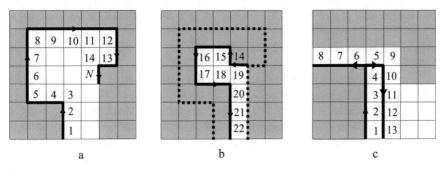

Fig. 7.4 Examples of reversible tracing: (**a**) left tracing trough the facets 1–14; (**b**) continued as right tracing through facets 15–22; (**c**) simultaneous left and right tracing

Fig. 7.5 A possible
extension of the fragment of
Fig. 7.4c; explanation in text

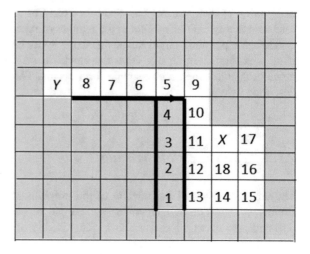

are examples of subsets of the opening frontier of the labeled region. The reason for
changing to opening frontiers is that the sequence of cracks of the closing frontier
with incident facets does not contain the facets incident to the „corners", such as e.g.,
facet 3 in Figure 7.4a. This was admissible for the algorithm of Sect. 7.1.1: A
missing facet was registered during the next turn of the spiral. However, this is not
admissible for the reversible tracing, as we shall see in what follows.

In most cases when a non-simple facet is met in the opening frontier, the tracing
can be continued in the reversed direction, as e.g., the sequence of Fig. 7.4a
continued in Fig. 7.4b. However, there are cases when directly after such a reversing
no simple facets can be found (Fig. 7.4c). Then the tracing must be continued
simultaneously in both directions. Two sequences of facets are then recorded: One
for the left and one for the right tracing. As soon as in one of these sequences a
simple facet is encountered, the other sequence is discarded. In this way the number
of non-simple facets in the resulting sequence is minimized.

In the example of Fig. 7.4c the left tracing goes on until non simple facet 5 is
reached. Since that location, two sequences have been recorded: 5, 6, 7, 8, etc. for the
left tracing and 5, 9, 10, 11 for the right tracing. At location 12 a simple facet is
found. Thus, the codes for the facets from 6, 7, 8 etc. are discarded. The sequence
5, 9, 10, 11 is transferred into the ultimate sequence and the tracing is continued from
facet 12.

The following question can arise: When will facets 6, 7, 8 etc. be picked up again
and included in the code? According to Theorem 2B ([4], p. 282), the frontier of
L always remains a 1-manifold, i.e., a closed sequence of points and cracks without
branching. Therefore, one can be sure that at some time later, the frontiers of these
facets will be reached by the tracing procedure and encoded (Fig. 7.5).

To make this assertion illustrative we can continue the image of Fig. 7.4c in two
different ways. If B is of genus 0, then the situation can appear as displayed in

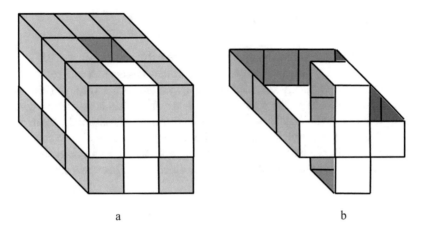

Fig. 7.6 An example of the facets of the rest sequence. (**a**) the simple facets of a torus are shaded, the non-simple ones are left white; (**b**) the set of non-simple facets of the same torus without the voxels

Fig. 7.5: Facets from 12 to 18 will be labeled since they are simple. After facet 18 has been labeled, facet X becomes simple and so do facets 11, 10, etc. until Y.

The ultimate sequence can contain (besides the rest sequence) some non-simple facets which are necessary to encode the path to further simple facets. However, the non-simple facets remain unlabeled. If the surface B is of genus 0, then the set of non-labeled facets *incident to the current frontier of L* becomes ever smaller until it is empty. Then the encoding of such a surface is finished.

The other possible way to extend the image of Fig. 7.4c is to regard it as a fragment of a surface of genus greater than 0. In this case the frontiers of facets 6, 7, 8 etc. will belong to the rest sequence as explained above and illustrated by Fig. 7.6. The facets will be included in the code as non-simple ones.

In the case of a surface of a genus greater than 0 (e.g., a torus) a closed sequence of non-simple facets arises at the end of the tracing procedure: The left and the right tracings are continued without finding a simple facet until they meet each other, while having opposite directions. The algorithm must be organized in such a way that the sequences contain *all non-simple facets*: If some facets at the starting point of the simultaneous tracing are missing, then they are lost forever. This is the reason for using the opening frontier rather than the closed one: The opening frontier is a *continuous* sequence of adjacent facets.

As in the method of Sect. 7.1.1, the non-simple facets contained in the rest sequence are labeled with „brute force ". The record of a facet in the rest sequence is changed as for a simple facet when the facet arrives in the rest sequence for the first time. After that, some records of non-simple facets remain at the end of the sequence. These are multiple records of facets which have been visited more than once. These records are employed for the calculation of the genus and then deleted.

7.1.3   Computer Experiments

The method of Sect. 7.1.2 has been successfully tested on many different 3D images. The first experiments were made with simple artificial subsets as a half-ball or a half-ball with a narrow vertically stretched parallelepiped attached at the upper side of the half-ball like a chimney. The surface of the latter object has a singularity where a vertical crack of the half-ball is simultaneously a crack of the chimney. More recent experiments have been conducted with tori of different size and with objects having up to nine tunnels, whose surfaces are of genus from 2 to 9. In all cases the surface was scanned and encoded completely. The 3D objects have been reconstructed from the code exactly.

7.1.4   Efficiency of Encoding

The facets of the sequence are encoded by 7 bits each as explained below. Thus, the code contains one byte per facet. The encoding can be made more economical by a factor of 2 by means of the difference crack-code as suggested in [5]. We have not realized this improvement to make the encoding and decoding as simple as possible.

The efficiency of encoding, measured as the ratio of the number of the facets in the surface to the number of elements in the ultimate sequence (plus one), is essentially better than that of the method of Sect. 7.1.1. It depends on the genus of the surface, as shown in Table 7.1 for the "worst case": Small objects with tunnels whose walls are only one voxel thick.

For larger objects, the efficiency is much better, as shown by the upper curve in Fig. 7.7. Thus, for example, the code of a parallelepiped of $38 \times 38 \times 19$ voxels with 9 tunnels parallel to the Z-axis is 8952 bytes long. The surface contains 7704 facets. The efficiency is equal to 0.861.

Figure 7.7 shows the efficiency of encoding small objects with the method of Sect. 7.1.2 (middle curve) as compared to that of Sect. 7.1.1 (lowest curve). The upper curve represents objects of $38 \times 38 \times 19$ voxels. These results can be compared with that of encoding a surface as a Euler circuit as described in [4], Sect. 13.3. In all our experiments the efficiency of the Euler circuit was about 1.4 facets per byte. (Unfortunately, another criterion of the efficiency was used in [4], Section 13.3, namely the ratio of number of bytes to the number of facets). It should be mentioned here that the method of Euler circuit is only applicable to a whole closed surface, whereas the method of spiral tracing can also be employed to encode

Table 7.1 Efficiency of the code as a function of the number of tunnels

Number of tunnels	0	1	2	3	4	5	6
Method of Sect. 7.1.1	1.0	0.485	0.336	0.231	0.179	0.147	0.125
Method of Sect. 7.1.2	1.0	0.842	0.685	0.624	0.619	0.608	0.557

any connected part of the surface. This possibility can be important e.g., for dissolving the surface into digital plane patches.

7.1.5 Examples of Codes

The codes of the facets are represented as octal numbers of three digits. The first digit specifies whether the facet is simple (digit $= 0$) or not (digit $= 1$). The second digit is the direction of the movement from a facet to the following crack; the third digit is the direction of the movement from the crack to the next facet. The directions are encoded as shown in Fig. 7.3 above.

1. A bar consisting of three voxels along the Y axis: 14 facets, 13 code elements.

   Coordinates of the first facet $= (3, 3, 4)$.
   The code:
   035, 040, 001, 011, 023, 035, 011, 020, 005, 013, 054, 044, 044.

2. A torus of 8 voxels; the axis of the tunnel is parallel to the Z axis: 32 facets, 37 code elements.

   Coordinates of the first facet $= (3, 3, 4)$.
   The code:
   035, 040, 000, 021, 015, 031, 023, 035, 011, 020, 000, 045, 051, 000, 144, 002, 011, 113, 124, 144, 144, 005, 053, 011, 032, 020, 011, 015, 033, 033, 054, 044, 044, 000, 100, 042, 021.

3. A B-shaped body with two tunnels parallel to the Z axis: 50 facets, 72 code elements:

   Coordinates of the first facet $= (3, 3, 4)$.
   The code:
   035, 040, 000, 021, 015, 031, 023, 035, 011, 020, 000, 045, 051, 000, 144, 002, 010, 051, 012, 000, 054, 044, 032, 020, 044, 133, 045, 000, 001, 011, 011, 023, 033, 033, 044, 044, 000, 015, 054, 033, 011, 032, 120, 111, 015, 033, 033, 054, 044, 044, 000, 100, 042, 121, 111, 135, 150, 144, 133, 133, 111, 111, 112, 100, 100, 124, 100, 100, 105, 144, 144, 053.

   The length of the code is always of the order of the number of the facets in the surface: In the case of genus zero the number of bytes in the code is equal or a little greater than the number of facets depending on the complexity of the shape of the surface. For surfaces with genus greater than zero the number of bytes is slightly greater than the number of facets, depending on the genus and on the size of the surface (compare Fig. 7.7).

   A student at the University of Applied Sciences Berlin C. Urbanek [6] has developed a computer program implementing this method and has conducted numerous experiments with encoding and reconstructing rather complicated bodies

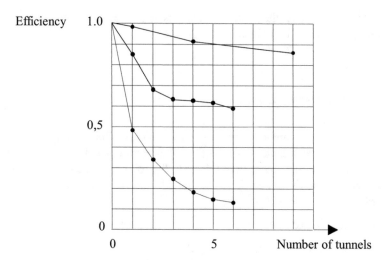

Fig. 7.7 Efficiency of encoding with different methods. Efficiency is the ratio of the number of the facets to the number of saved bytes plus 1

with many tunnels of different orientations. He has only referenced the publication [7] and has accomplished his work without having consulted the author.

7.1.6   Conclusion

A new method is suggested of representing a surface in the 3D space as a single continuous sequence of facets. According to this method the surface is encoded as a single sequence of mutually adjacent facets. Each facet is encoded by one byte. The code of the surface of a three-dimensional object takes much less memory space than the raster representation of the object. The object can be reconstructed from the code exactly. Surfaces of a genus greater that zero (e.g., that of a torus) can also be encoded by a single continuous sequence. The algorithm recognizes the genus of the surface. The method is well suited for dissolving a given surface into patches of digital planes, which is its advantage over known methods.

7.2   Algorithm CORB_3D for Traversing and Encoding Surfaces

We have described in Chap. 4 an algorithm CORB for tracing and encoding boundaries of homogeneous connected subsets of an 2D image. (In a homogeneous subset of a two-dimensional image all pixels have one and the same color). Now we

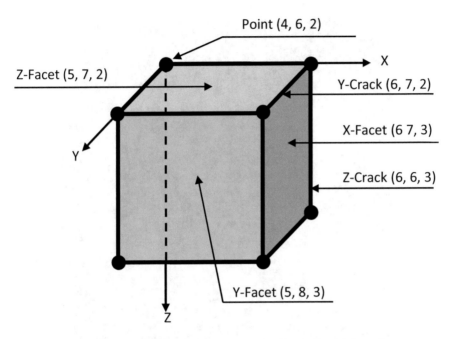

Fig. 7.8 A voxel (5, 7, 3) with the cells of its closure and with combinatorial coordinates

describe a similar algorithm CORB_3D for tracing and encoding surfaces in 3D images. A surface is the boundary of a connected homogeneous subset of voxels. According to the topological definition, a boundary of a subset S of the space R is the subset of cells whose smallest neighborhood crosses S and its complement $R - S$. As stated in Sect 2.1 and illustrated in Fig. 2.5, the smallest neighborhood of a three-dimensional cell or a voxel in a three-dimensional complex consists of a single cell, namely the voxel itself. Therefore, the smallest neighborhood of a voxel cannot interest both the subset S and its complement $R - S$. Thus, a voxel cannot belong to the boundary. The boundary of a subset of a three-dimensional complex consists of cells of dimension less than 3. These are the two-dimensional cells called facets, the one-dimensional cells called cracks and the zero-dimensional cell called points. If we represent voxels as small cubes, then facets are the sides of the cubes, cracks are edges of the cubes and points are the corners of the cubes. Figure 7.8 shows a voxel with incident cells of lower dimensions and their combinatorial coordinates.

Figure 7.9 shows an example of the surface of a digital ball with radius = 5.

We describe below the algorithm "CORB_3D". It contains the sub-algorithms "Search3D" and "Trace3D".

The aim of the algorithm "CORB_3D" is to traverse and encode each component of the boundary of each homogeneous subset of a 3D image. An especial property of the algorithm "CORB_3D" is like the property of the 2D algorithm: the traversing of

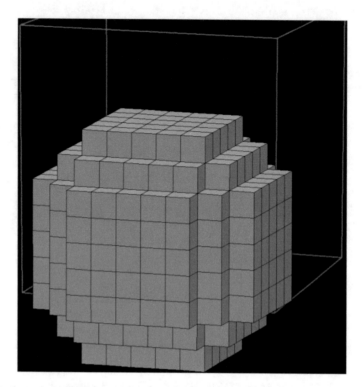

Fig. 7.9 Example of the surface of a digital ball with radius = 5

a boundary component chooses one of two densities on the sides of the boundary as the "foreground density" and considers all other densities as the background. The traversing is performed in such a way that the foreground density remains always at the same side of the traversing. Thus, the image is from the view of traversing a *binary image* with two colors: foreground and background while background is any density different from the foreground density. Correspondingly the boundary becomes a closed surface, as all boundaries in a binary 3D image. This is an important advantage of the algorithm "CORB_3D", and it makes the algorithm so simple.

If a region R_1 of the image contains a hole, which is another region R_2 of the image, then the boundary of R_2 will be traced and encoded twice: first as a component of the boundary of R_1 with the density of R_1 as foreground density and then as the outer boundary of the hole R_2 with the density of R_2 as foreground density. In a binary image being regarded as a cell complex each component of the boundary is a set of points, cracks, and facets.

The algorithm finds a boundary facet, i.e., a facet incident to two voxels with different densities. The algorithm checks whether this facet was labeled as already visited. If it was not visited, then the algorithm fixes one of these densities as the foreground density and starts the traversing procedure which must traverse a boundary component of a subset having the foreground density. The starting crack of the traversing is one of the four cracks incident to the found facet. The algorithm saves for each component of the surface (which we call a "bubble") the coordinates of all facets of the component. The facets which are visited during the traversing must be labeled as "already visited" so that no of these labeled facets can be used to start the traversing procedure again. The coordinates of all facets of the boundary saved for all boundary components are sufficient for an exact reconstruction of the original image in the way which will be explained below.

To reconstruct the original image from the codes of the boundaries it is also necessary to encode the outer boundary of the whole image. For this purpose, we consider a virtual outer subset R_0 having a non-existing density, e.g., -1. Then another subset R_1 is regarded as contained in R_0. The components of the boundaries of the regions of the whole image compose a structure of an *inclusion tree* which shows which components are contained in the interior of some other component. The algorithm makes the field *List* containing the coordinates of all boundary facets and a structure *Bubble* for each component of the boundary. The structure *Bubble* contains the index *First* which is the index of the first facet of *Bubble* in the field *List* and the index *Last* which is the index of the last facet of *Bubble* in the field *List*. The structure *Bubble* also contains the variable *Density* which is the value of the densities contained in the voxels of the traced subset.

We describe below the algorithms for traversing boundaries in 3D images and for restoring the original image from the fields *List* and *Bubble*. The most important methods (subroutines) are the sub-algorithms "Search3D", "Traverse", and "Restore3D". The algorithm "Search3D" scans the image plane by plane, row by row, and looks for a pair of subsequent voxels with different densities. If the facet lying between these two voxels was not already visited, the method fixes the density of the first of these two voxels and calls the algorithm "Traverse" with the fixed density as the foreground density and the coordinates of the starting facet as arguments.

In the following algorithm "Search3D" the notation "Voxel[width, height, depth]" stands for a three-dimensional array whose elements contain densities. The values "width", "height" and "depth" are the numbers of columns, rows and planes in the field "Voxel[,,]" respectively. Similarly, "Label[width + 1, height, depth]" is a global three-dimensional array of the size (2*width*height*depth). The size of "Label" is increased since it is necessary to label facets at the sides of "Voxel". The constant "Out" is the virtual, non-existing density of voxels outside the image.

Other than in the two-dimensional case, the boundary of a three-dimensional subset (a body) is described not by loops but rather by "bubbles" looking like spheres.

The variable *nBubble* of the following algorithm "Search3D" defines the number of already detected bubbles, and the variable *nList* defines the number of already detected facets in the bubbles. These variables must be saved in such a way that they can be accessible to the algorithm "Restore3D" described below. The values of these variables are transmitted to the algorithm "Traverse" as parameters.

Algorithm Search3D(Vector Sz, Image[Sz.X, Sz.Y, Sz.Z]):

Step 1: Declare local variables Density, value, oldValue, Out,
                          sx, sy, sz, (standard coordinates)
                          x, y, and z. (combinatorial coordinates)
Step 2: Declare three-dimensional vectors Facet and Voxel.
Step 3: Set Out ← -1.
Step 4: Initialize global variables nBubble ← 0; nList ← 0. (Indices of Bubble and list of facets)
Step 5: Set x ← 1.
Step 6: Repeat the steps until x is less than 2*Sz.X:
    6.1: Set sx ← x / 2, Voxel.X ← x, and y ← 1.
    6.2: Repeat the steps until y is less than 2*Sz.Y:
    6.2.1: Set sy ← y / 2, Voxel.Y ← y, and z ← 2*Sz.Y.
    6.2.2: Repeat the steps until z is greater or equal to 1:
    6.2.2.1: Set sz ← z / 2, Voxel.Z ← z and
              value ← Image[Voxel.X, Voxel.Y, Voxel.Z].
    6.2.2.2: If sz + 1 is less than Sz.Z, then
                          set val_old ← Image[sx, sy, kz + 1].
    6.2.2.3: If value is not equal to val_old, then:
        Begin 6.2.2.3
          Set Facet ← Voxel and Facet.Z ← Facet.Z +1. (Start Facet)
          If the value of GetLabel(Facet, 'L') equals 0 AND
                BoundaryPixelN(Facet, Norm) equals TRUE, then:
          Begin If
            Density ← val_old.
            If GetLabel(Facet, 'L') equals 0, then:
            Begin
              Call Traverse(Facet, Density, nBubble, nList).
              Set nBubble ← nBubble + 1.
            End
          End If
        End 6.2.23
      6.2.2.4: Set z ← z – 1.
      6.2.3: Set y ← y + 1.
    6.3: Set x ← x + 1.
Stop.

The algorithm "Traverse" runs along the boundary and makes records into the structure *Bubble[nBubble]* of the actual bubble and into the array *List*, where the coordinates of the boundary facets are stored. The structure *Bubble* contains the integer foreground density assigned to the bubble and the indices "First" and "Last" of the records of the first and last facets of the bubble in the array *List*.

Algorithm Traverse(iVect F_Start, int Density, int nBubble, int nList):

Step 1: Declare the variable Q as a three-dimensional vector.
Step 2: Push the variable F_Start into the stack.
Step 3: Call utility algorithm PutLabel(F_Start, 'S')   (to label F_Start
                                                                      as already pushed to the stack.)
Step 4: Set the variable Bubble[nBubble].First ← nList.
Step 5: Repeat if the stack is not empty:
   5.1: Pop the variable Q from the stack.
   5.2: Set the variable List[nList] ← Q.
   5.3: Set the variable nList ← nList + 1.
   5.4: If the coordinate Q.Z is even, then:
    Begin
      Declare a new three-dimensional vector Vox with the coordinates
                                          Q.X, Q.Y, Q.Z + 1.
      Calculate the value ValVox of Vox by calling
                            the utility algorithm Value2(Vox, 3):
                  ValVox ← Value2(Vox, 3).
      If ValVox is equal to Density, then:
                            Label Q as already saved in List.
    End
   5.5: Call the utility algorithm AllNeighbors(Q, Density).
   End Step 5.
Step 6: Set Bubble[nBubble].Last ← nList – 1.
Step 7: Return nList.
Stop.

We describe now the most important utility algorithm "AllNeighbors(iVect F, int Density)". This algorithm calculates all facets adjacent to the given facet "F". These facets are either adjacent to "F" through the four cracks incident to "F" or adjacent to "F" through the singular corner points of "F". A corner point of "F" is singular if it is incident with more than four boundary cracks.

Algorithm AllNeighbors(iVect F, int Density):

Step 1: Declare three-dimensional vectors Crack, Fnew, and Norm.
Step 2: Calculate the vector Norm by means of utility algorithm GetNorm:
             Norm ← GetNorm(F, Density):
Step 3: Declare the vector field Point[4].
Step 4: Declare and initialize the constant vector fields StepX[4], StepY[4], StepZ[4]:
  StepX[0] ← (0, 1, 0), StepX[1] ← (0, 0, 1), StepX[2] ← (0, -1, 0), StepX[3] ← (0, 0, -1).
  StepY[0] ← (1, 0, 0), StepY[1] ← (0, 0, 1), StepY[2] ← (-1, 0, 0), StepY[3] ← (0, 0, -1).
  StepZ[0] ← (1, 0, 0), StepZ[1] ← (0, 1, 0), StepZ[2] ← (-1, 0, 0), StepZ[3] ← (0, -1, 0).
Step 5: If Norm.Y is negative, then label F with label "N".
Step 6: If F.X is even, then: (F is an X-facette)
      Begin
      Set I ← 0.
      Repeat until I is less than 4:
      Begin
      Set Crack ← F + StepX[I].
      Call utility algorithm Next_Face(Density, F, Crack, ref Fnew).
                  ("ref" means Fnew will be changed by Next_Face)
     If Fnew is a boundary facet and GetLabel(Fnew, 'S') equals 0, then:
      Begin
      Label F with label 'S'.
      Push Fnew to the stack.
      End
      Set I ← I + 1.
      End
Step 7: If F.Y is even, then: (F is a Y-facette)
      Begin
      Set I ← 0.
      Repeat until I is less than 4:
      Begin
      Set Crack ← F + StepY[I].
      Call utility algorithm Next_Face(Density, F, Crack, ref Fnew).
                       ("ref" means Fnew will be changed)
     If Fnew is a boundary facet and GetLabel(Fnew, 'S') equals 0, then:
      Begin
      Label F with label 'S'.
      Push Fnew to the stack.
      End
      Set I ← I + 1.
      End
Step 8: If F.Z is even, then: (F is a Z-facette)
      Begin
      Set I ← 0.
      Repeat until I is less than 4:
      Begin
      Set Crack ← F + StepZ[I].
      Call utility algorithm Next_Face(Density, F, Crack, ref Fnew).
                       ("ref" means Fnew will be changed)

If Fnew is a boundary facet and GetLabel(Fnew, 'S') equals 0, then:
    Begin
    Label F with label 'S'.
    Push Fnew to the stack.
    End
    Set I ← I + 1.
    End
Step 9: Call the utility algorithm IsoSingPoints(F, Point, Density) and
        set the number of singular points nPoint← result of IsoSingPoints().
Step 10: Declare a field of 5 cracks Crack[5].
Step 11: Declare a field of 6 facets Fn[6].
Step 12: Set the variable I← 0.
Step 13: Repeat until I is less than 4:
        Begin 13
        Set Crack[1] ← F and Crack[2] ← F.
        If F.X is even, then set Crack[1].Y ← Point[I].Y and Crack[2].Z ← Point[I].Z.
        If F.Y is even, then set Crack[1].X ← Point[I].X and Crack[2].Z ← Point[I].Z.
        If F.Z is even, then set Crack[1].Y ← Point[I].Y and Crack[2].X ← Point[I].X.C
        Set Crack[3] ← Point[I] + Point[I] – Crack[1].
        Set Crack[4] ← Point[I] + Point[I] – Crack[2].
        Set Fn[1] ← Crack[3] + Norm.
        If Fn[1] is a boundary facet and GetLabel(Fn[1], 'S') equals 0, then:
            Label Fn[1] with 'S' and push Fn[1] to the stack.
        Set Fn[2] ← Crack[3] - Norm.
        If Fn[2] is a boundary facet and GetLabel(Fn[2], 'S') equals 0, then:
            Label Fn[2] with 'S' and push Fn[2] to the stack.
        Set Fn[3] ← Crack[4] + Norm.
        If Fn[3] is a boundary facet and GetLabel(Fn[3], 'S') equals 0, then:
            Label Fn[3] with 'S' and push Fn[3] to the stack.

        Set Fn[4] ← Crack[4] - Norm.
        If Fn[4] is a boundary facet and GetLabel(Fn[4], 'S') equals 0, then:
            Label Fn[4] with 'S' and push Fn[4] to the stack.

        Set Fn[5] ← Point[I] + Point[I] – F.
        If Fn[5] is a boundary facet and GetLabel(Fn[5], 'S') equals 0, then:
            Label Fn[5] with 'S' and push Fn[5] to the stack.
        Set I ← I + 1.
        End.
Stop.

We have prepared an artificial 3D image of the size of 100*100*100 voxels containing a parallelepiped of the size of 50*50*50 voxels with the density equal to 64 and an adjacent cylinder with a horizontal axis going through the central point of the rectangle perpendicular to one side of the parallelopiped with the density of 128. Then we have started the algorithm "Search3D" for this artificial image. It has produced three bubbles with parameters shown in Table 7.2.

Table 7.2 Parameters of the bubbles 0 to 2

Index	Density	Number facets	First facet
0	0	22'156	(33, 101,150)
1	64	18'576	(33,101,150)
2	128	4'928	(119,151,94)

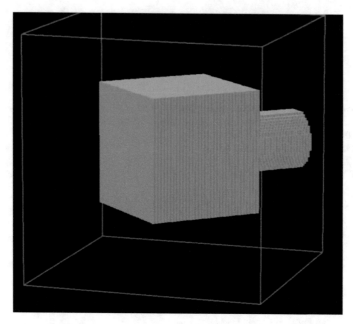

Fig. 7.10 Central projection of the body in an artificial 3D image

Central projections of the bubbles are shown in Fig. 7.10.

We have tested our algorithms also with true medical 3D images produced by computer tomography (CT).

We have taken from the Internet a CT image and transformed it by means of the standard viewer "MicroDicom" into a series of JPG images. Our special algorithm has read these images one after another and has copied the values in their pixels to the values in the voxels of a 3D image. Each two-dimensional subset of this 3D image corresponding to fixed value of the coordinate Z was a copy of a single JPG image.

As in the 2D case, a 3D image can contain a great number of small homogeneous subsets, some of them consisting of a single voxel: all voxels adjacent to this voxel have densities not equal to the density of the single voxel. The surfaces of such small homogeneous subsets are of no interest for the analysis of the contents of the image. Therefore, it is rational to transform the 3D image by reducing the number of different values of the density. Due to this transformation some small homogeneous subsets with similar densities will merge composing greater subsets. In the case of the 2D images we have used a similar transformation while transforming a color image containing up to about 16 million of colors to an indexed image with an essentially smaller number of colors.

One possible way of reducing the number of different values of the density consists in calculation the histogram of the densities, to smooth the histogram and to find local minima of the values of the smoothed histogram. The local minima are thresholds for the quantization of the densities, while the first threshold should be set to zero and the last threshold should be set to 255. All values of the density of the

Fig. 7.11 Central
projection of one component
of 3D image of the computer
tomography (angiography)

original image lying between two adjacent thresholds T and S should be replaced by
the value $(T + S)/2$.

The algorithm "CORB_3D" calculates the list of all boundary facets and a
structure "Bubble" for each connected component of the boundary. A bubble
contains the values "First" and "Last" being the indices of the begin and the end
of the sub-list (a part of the list) corresponding to the bubble. Figure 7.11 shows the
central projection of the boundary of an object in the 3D image of the computer
tomography (angiography) of the blood vessels of a human body.

7.2.1 Properties of the Bubbles

The boundary of a subset S of the image I is the set of all facets, cracks and points
whose smallest open neighborhood (SON) crosses both S and its complement $I - S$.

1. If the subset S is a connected set of all voxels with one and the same density D,
   then the boundary of S is a closed surface called bubble. This is true because
   from the site of D the image is a binary one: D is foreground and "not D" is
   background. It is well known that each boundary in a binary 3D image is a set of
   closed surfaces. Each bubble in the boundary of S obtains D as its proper
   foreground density which we call in what follows the "*for_density*".

2. If a bubble contains its *for_density* in its interior, then it is called a *positive*, otherwise a *negative* bubble.

3. The boundary of the whole 3D image can also be considered as a bubble with the not existing *for_density* "Out" or − 1. This bubble is negative since it does not contain its *for_density*.

4. It is possible to calculate a inclusion tree for all bubbles of an image. The bubbles of an image are ordered: each bubble obtains a number according to the order of the calculation. A bubble obtains in the inclusion tree three labels: *father, son,* and *brother*. The father of the bubble L is the number of the bubble containing L immediately. Immediately means that there is no other bubble that lies in the father bubble and contains L in its interior.

   The son of the bubble L is the eldest son of L which means that son is the smallest number among the numbers of the bubbles immediately contained in L.

   Brother of the bubble LB is the eldest brother of LB. A brother can be a brother of the son or the brother of another brother. This means that the brother of the bubble LB is the smallest number among the numbers of the bubbles immediately contained in the bubble "father of LB" and being greater than the number of the bubble LB.

5. Each negative bubble LN contains several but at least one positive bubble. These positive bubbles contain all in the bubble LN contained voxels so that the sum of the volumes of these positive bubbles is equal to the volume of LN.

6. A negative bubble LN has no immediately contained voxels. This means that if a voxel is contained in LN, then it is also contained in a positive bubble immediately contained in LN.

7. The negative and positive bubbles of an image form layers in the inclusion tree: the layer 0 is the singe negative bubble "Bubble[0]" containing the whole image. The layer 1 is formed by all positive bubbles which are immediately contained in "Bubble[0]". The layer 2 is formed by all negative bubbles which are immediately contained in the positive bubbles of the layer 1 etc. General: The layer $(n + 1)$ is formed by all positive bubbles immediately contained in the negative bubbles of the layer (n). The layer $(n + 2)$ is formed by all negative bubbles which are immediately contained in the positive bubbles of the layer $(n + 1)$.

8. A negative bubble can be immediately contained only in a positive bubble.

9. A positive bubble can be immediately contained only in a negative bubble.

10. A connected set of voxels of a single density can be immediately contained *only* in a positive bubble LP. The volume of such a set is equal to the volume of the positive bubble LP minus the sum of the volumes of all the negative bubbles immediately contained in this positive bubble LP.

11. If the father of the bubble L has the same *for_density* as the bubble L, then L is a negative bubble. If, however, the father of the bubble L has a *for_density* different from the *for_density* of L, then L is a positive bubble.

7.3   Theory of Digital Plane Patches

(Section 7.3 was quoted from the author's book [4] with the permission of the publisher).

In Sect. 5.1.1 we have defined a digital half-space as a region containing all ground cells of a Cartesian space, whose coordinates satisfy a linear inequality. Now we can use this definition to introduce the notion of a digital plane and of a digital plane patch. As in the 2D case, we will suggest considering the frontier rather than the boundary of a half-space since the boundary contains some subsets of the boundary of the space. Remember that the boundary of a subset $T \subseteq S$ of a finite n-dimensional Cartesian space S is the closure of the set of all $(n-1)$-cells each of which is bounding exactly one n-cell of T.

Definition DPP A *digital plane patch* (DPP) is any connected subset of the frontier of a three-dimensional half-space.

According to the above Definition, a DPP is the subset of the surface of a 3D complex whose voxel coordinates satisfy a linear inequality. It consists of 2-cells and their boundaries. A DPP contains no voxels. One may find in the literature either DPPs consisting of 2-cells [8] or DPPs consisting of voxels [9]. There is no need to consider "thick" surfaces composed of voxels. Many authors came to the idea that voxels are small cubes rather than points of a 3D grid (or points in Z^3). This idea coincides with the idea of cell complexes: From the point of view of cell complexes surfaces are sets of two-dimensional cells or facets, which are sides of the cubes representing the voxels.

Let us consider the inequalities defining a DPP.

Definition COPL A cell d is said to be *strictly coplanar* with three other cells a, b and c if the following condition holds:

$$D = \begin{vmatrix} x_d - x_a & y_d - y_a & z_d - z_a \\ x_b - x_a & y_b - y_a & z_b - z_a \\ x_c - x_a & y_c - y_a & z_c - z_a \end{vmatrix} = 0;$$

where D is the 3×3 determinant composed of the differences of the coordinates of the said cells.

The cell d is said to lie *to the positive side* of the ordered triple of the a, b and c if $D > 0$.

It lies *to the negative side* of the cells a, b and c if $D < 0$.

In other words, the cell d lies to the positive side of (a, b, c) if the three vectors d-a, b-a, c-a compose a system whose orientation is the same as that of the three basis vectors x, y, z of the used coordinate system.

Exactly as in the 2D case, the above definitions and inequalities are applicable without changes for both the mathematical coordinate system which is a right-hand one and for the left-hand coordinate systems used in computer graphics. The notions of right and left do change when getting over from one system to the other; the notions of positive and negative do not.

7.3.1   Properties of Digital Plane Patches

We consider here digital plane patches (DPPs) rather than digital planes because in practice one has always to do with finite subsets of a digital space. Such a subset can only contain a subset of a plane, i.e., a plane patch. We shall investigate the conditions satisfied by the *combinatorial coordinates* of cells of a DPP. These conditions are different from those in standard coordinates since the parity of combinatorial coordinates of a cell depends on its dimension and on the orientation of one- and two-dimensional cells.

To investigate the properties of DPPs it is easier and more comprehensible to consider at first the 3-cells (voxels) incident to the cells of a DPP rather than the cells of the DPP themselves. Consider the boundary of a strongly connected subset of a three-dimensional Cartesian space. Let us call such a subset a "body". It is a two-dimensional quasi-manifold [4], page 155.

Definition DS (digital surface) A connected subset of a two-dimensional quasi-manifold is called a *digital surface K* in a three-dimensional space.

It is possible to assign an orientation to the digital surface K and thus to the 2-cells of K, e.g., in such a way that the positive normal to each 2-cell of K points to the outside of the body. Suppose that K does not intersect the boundary of the space. Then each 2-cell F of K is incident to exactly two voxels. One of them lies to the positive side, i.e., to the side which the positive normal is pointing to. The other voxel lies to the negative side of F. In what follows we shall call the 2-cells *facets*.

Definition PP A voxel P, which is incident to an oriented facet F of the surface K and lies to the positive side of F is called the *positive voxel* of K. Similarly, the incident voxel lying to the negative side of F is called the *negative voxel* of K.

The set of all positive voxels of K will be called the *positive voxel set* of K and denoted by $SP(K)$. The set of all negative voxels of K will be called the *negative voxel set* of K and denoted by $SN(K)$.

If K is a DPP, then, according to Definition DPP, there is a linear form $H(x, y, z)$ such that $H(x, y, z) \geq 0$ for all positive voxels of K and $H(x, y, z) < 0$ for the negative ones. There also exits another linear form $H(x, y, z)$ (differing from $H(x, y, z)$ by a constant) such that $H(x, y, z) > 0$ for all positive voxels of K and $H(x, y, z) \leq 0$ for the negative ones. We shall call both linear forms $H(x, y, z)$ and $H(x, y, z)$ the *separating* ones.

The properties of a DPP, which we are interested in, and which are important for the recognition of a DPP are known from the literature [9] during many years. However, we shall deduce these properties anew since most of the publications consider *thick digital planes in standard coordinates, which are sets of voxels,* and we are interested in *boundary planes in combinatorial coordinates*. Also, the investigation by means of combinatorial coordinates is simpler than that by means of standard coordinates. The most part of our results is applicable for both standard and combinatorial coordinates due to denoting the minimum distance between two voxels by e which is equal to 1 in standard coordinates and to 2 in combinatorial ones.

We choose the coefficients of the separating form as the components of a vector being orthogonal to the oriented DPP under consideration. We choose the signs of the coefficients so that the form be positive for voxels lying to the positive side of the oriented facets.

Definition SSF   The separating linear form $H(x, y, z) = a \cdot x + b \cdot y + c \cdot z + r$ of a DPP D is called the *standard separating form* (SSF) of D if it satisfies the following conditions:

$H(x, y, z) \geq 0$ for all positive voxels of D, $H(x, y, z) < 0$ for all negative ones;

The coefficients a, b, and c of $H(x, y, z)$ are integers while their greatest common divider is equal to 1;

The coefficients a, b, c are related to the coordinates of voxels incident to D in such a way that one of the following conditions is fulfilled:

(a) there are at least three positive voxels P_1, P_2 and P_3 of D at which $H(x, y, z)$ takes its minimum value *with respect to* all positive voxels of D;
(b) there are at least three negative voxels N_1, N_2 and N_3 of D at which $H(x\,y, z)$ takes its maximum value *with respect to* all negative voxels of D;
(c) there are at least two positive voxels P_1 and P_2 of D at which $H(x, y, z)$ takes its minimum value and at least two negative voxels N_1 and N_2 of D at which $H(x, y, z)$ takes its maximum value.

Definition BS   The set of positive voxels of a DPP D at which the SSF of D takes its minimum value *with respect to* all positive voxels of D is called the *positive base* of D. Similarly, the set of negative voxels of a DPP D at which the SSF of D takes its maximum value with respect to all negative voxels of D is called the *negative base* of D.

We shall formulate and prove in what follows a set of theorems concerning the properties of a DPP. These theorems are necessary to prove that the suggested algorithm for recognizing a DPP is correct.

First, we shall show that any DPP of a finite size (of at least three facets) possesses the standard separating form.

Theorem SSF   For any DPP D (of at least three facets) in a Cartesian 3D-space there exists the standard separating form. The absolute values of the coefficients a, b and c are not greater than the greatest dimension of the space.

Proof   Consider the convex hulls of the set of all positive and of the set of all negative voxels (being considered as points with the corresponding coordinates rather than as cubes). Let us denote them by CH^+ and CH^- correspondingly. They do not overlap; otherwise, there would be no possibility to separate the sets by a linear form and hence D could not be a DPP. The line segment L corresponding to the shortest distance between the convex hulls CH^+ and CH^- has at each end either a voxel, or an edge of a hull orthogonal to L, or a face of a hull orthogonal to L. Consider the linear form $H(x, y, z) = a{\cdot}x + b{\cdot}y + c{\cdot}z + r$ whose coefficients (a, b, c) compose a vector collinear with L and directed from CH^- towards CH^+, while the coefficient r has such a value that H is equal to 0 at the endpoint of L lying in CH^+. It is obviously the standard separating form. Its coefficients (a, b, c) are integers since they compose a solution of a system of homogeneous linear equations with integer coefficients. The coefficients (a, b, c) may be made mutually prime by the division by their greatest common divisor which does not violate the condition that the vector (a, b, c) is collinear with L.

Theorem 3O (3 orientations) An oriented DPP D in a Cartesian complex contains oriented facets (2-cells) of at most three orientations.

The proof of the Theorem 3O is almost identical to that of Theorem TWD of Sect. 5.1.2 and is omitted.

We shall demonstrate that the facets of different orientations in a DPP compose a certain pattern in which one of the orientations occurs singularly and another one semi-singularly as explained below. Two facets are called *strongly adjacent* if there is a crack bounding both.

Consider a facet F of D not incident to the boundary of D. Then it has exactly four strongly adjacent facets. Let us denote the set of facets of D strongly adjacent to F by S. An orientation O of F is called *singular* in D if all facets of S have orientations different from O. An orientation O of a facet F in a DPP D is called *semi-singular* in D if there are in S two facets, which are not adjacent to each other and have orientations different from O. These two facets are separated from F by cracks which are parallel to each other. Otherwise, the orientation O is called the *main orientation* of D (Fig. 7.12).

We shall say in what follows that a facet has the orientation X_i if its normal is parallel to the coordinate axis X_i. Let us rewrite the separating form as

$$H(x, y, z) = \sum a_i x_i + r; \tag{7.2}$$

and denote the indices of the coefficients in such a way that the following inequalities hold:

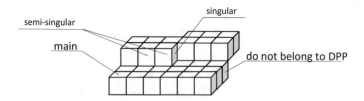

Fig. 7.12 A DPP and its singular and semi-singular orientations

Fig. 7.13 To the proof of
theorem SING1

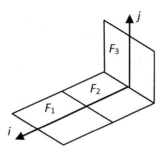

$$| a_{\mathrm{Imin}} | < | a_{Imid} | < | a_{\mathrm{Imax}} | .\qquad(7.3)$$

Theorem SING1 In a DPP with the separating form H according to (7.2) and (7.3) the orientation X_{Imin} is singular and the orientation X_{Imid} is semi-singular.

Proof: The Idea of the Proof If in a DPP D there are two parallel adjacent facets F_1 and F_2 whose i-th coordinates differ by 2 and another facet F_3 adjacent to F_2 with the orientation X_i whose j-th coordinate differs from those of F_1 and F_2 by 1, then $|a_i| \geq |a_j|$ (Fig. 7.13).

Corollary VF (value at a facet) Let $H(u) = H(x, y, z) = a\cdot x + b\cdot y + c\cdot z + r$ be the SSF of the DPP D and F be a facet of D having a non-singular orientation, while x, y and z are the combinatorial coordinates of a cell u. Then $H(F) = H(N) + |a_{Imax}| = H(P) - |a_{Imax}|$, where N is the negative and P the positive voxel incident to F. If F has the singular orientation, then $H(F) = H(N) + |a_{Imin}| = H(P) - |a_{Imin}|$. If F has the semi-singular orientation, then $H(F) = H(N) + |a_{Imid}| = H(P) - |a_{Imid}|$.

Proof Let the facet F have the orientation X_i and P be a voxel incident to F. Then the combinatorial coordinate x_i of F differs from that of P by 1 and the two other coordinates of F are equal to that of P. Therefore, the values $H(F)$ and $H(P)$ differ by $|a_i|$. The same is true for the values $H(F)$ and $H(N)$ where N is another voxel incident to F. If P is a positive and N the negative voxel of D, then according to the definition SSF $H(P) > H(N)$. Since $H()$ is linear, the value $H(F)$ lies between $H(P)$ and $H(N)$:

$$H(F) = H(N) + | a_i |= H(P) - | a_i | .\qquad(7.4)$$

According to theorem SING1 the non-singular facets of a DPP have the orientation X_{Imax}. Thus, in this case $a_i = a_{Imax}$. Similarly, in the case of semi-singular facets the orientation is X_{Imid} and $a_i = a_{Imid}$ and in the case of singular facets the orientation is X_{Imin} and $a_i = a_{Imin}$.

Theorem INC1 A positive voxel of a DPP D, which is not incident to the boundary of D, is incident to a facet of the main orientation.

Proof Consider a positive voxel V, which is not incident to the boundary of D. Then it is incident to a facet F of D, which is also not incident to the boundary of D. Thus, F is adjacent to exactly four other facets composing the set S. If F has the main orientation, then we are done. Let the orientation of F be O. If O is semi-singular, then there are in S two non-adjacent facets F_1 and F_2 with an orientation O^* different from O. This orientation must be the main one. Otherwise, it could only be singular, but in this case O would be also singular. It may be shown that in this case all three orientations are singular.

If O^* is the main orientation, then one of the facets F_1 and F_2 is incident to V and we are done. In the remaining case O is singular. Then thereare in S two pairs of non-adjacent facets each pair having an orientation different from O. Thus, one of these two orientations is the main one and again one of the facets F_1 and F_2 is incident to V. We have considered all possible cases.

Theorem VAL Let $H(x_1, x_2, x_3) = \sum a_i x_i + r$ be the standard separating form of the DPP D and (x_1, x_2, x_3) are combinatorial or standard coordinates of a positive voxel of D. Then the values of $H(x_1, x_2, x_3)$ for all positive voxels of D lie in the closed interval $[0, e \cdot |a_{\mathrm{Imax}}| - 1]$, where $e = 2$ for combinatorial coordinates and $e = 1$ for standard coordinates. The values of $H(x_1, x_2, x_3)$ for all negative voxels of D lie in the closed interval $[-e \cdot |a_{\mathrm{Imax}}|, -e]$.

Proof The standard separating form H separates the voxels in such a way that $H(x_1, x_2, x_3) \geq 0$ for all positive and $H(x_1, x_2, x_3) < 0$ for all negative voxels. Let P be a positive voxel of D and let P be incident to a facet F_i of orientation X_i. F_i is incident to a negative voxel N of D as each facet of a DPP does per Definition PV (if the facet does not lie in the boundary of the space). Thus, the vector $P - N = \pm e \cdot n_i$ where n_i is the i-th unit base vector of the coordinate system. The sign should be chosen in such a way that $H(P) > H(N)$. Therefore $H(P) = H(N) + e \cdot |a_i|$. Since the coefficients of a SSF are integers, $H(N)$ is also an integer. Thus from $H(N) < 0$ follows $H(N) \leq - e$ and $H(P) \leq e \cdot (|a_i| - 1) \leq e \cdot (|a_{\mathrm{Imax}}| - 1)$. Similarly, it follows from $H(P) \geq 0$ that $H(N) \leq 0 - e \cdot |a_{\mathrm{Imax}}|$ and $-e \leq H(N) \leq - e \cdot |a_{\mathrm{Imax}}|$.

7.3.2   The Problem of the Segmentation of Surfaces into DPPs

This problem is of great practical importance since its solution promises an economical and precise encoding of 3D scenes.

Problem Statement
Given: a surface in a 3D space, e.g., the boundary of a connected subset.

Find: the minimum number of DPPs representing the surface in such a way that it is possible to reconstruct the surface, at least with a predefined precision, from the code of the DPPs.

The problem consists of two partial problems:

1. *Recognition:* Given a set of facets decide whether it is a DPP or not.
2. *Choice*: Given a surface S and a subset of facets, which is known to be a DPP, decide which facet of S should be appended to the subset to achieve that the number of the DPPs in the segmentation of S be minimal.

7.3.3   The Partial Problem of the Recognition of a DPP

Similarly, as in the case of the DSS (Sect. 5.1), a surface S specifies two sets of voxels incident to the facets of S: a positive and a negative set. The voxels of the positive set lie outside of the body whose boundary is S, that of the negative set lie inside. If a subset T^2 of facets of S is a DPP, then it lies in the frontier of a half-space whose voxels satisfy a linear inequality. The inequality separates the positive voxels of T^2 from its negative voxels.

To decide whether T^2 is a DPP it is sufficient to solve the following system of $2 \cdot N$ linear inequalities in the components H_k, $k = 1, 2, 3$; of a 3D vector H and a scalar value C where N is the number of facets in T^2.

$$
\begin{aligned}
&\textstyle\sum_k H_k \cdot V_k^+(F_i) - C \geq 0; \\
&\text{with } F_i \in T^2; i = 1 \ldots N; \\
&\textstyle\sum_k H_k \cdot V_k^-(F_i) - C < 0;
\end{aligned}
\tag{7.5}
$$

$V_k^+(F_i)$ stays for the kth combinatorial coordinate of the positive voxel incident to the face F_i. Similarly, $V_k^-(F_i)$ is the kth combinatorial coordinate of the negative voxel incident to F_i. The vector H is the normal to a plane separating all positive voxels from the negative ones, while C specifies the distance of the plane from the coordinate origin.

It is possible to solve the problem by a fast method like that of recognizing a DSS (Chap. 5). The corresponding algorithm and the related theory are, however, rather complicated. Their presentation here is impossible because of the page limit. We describe a rather simple algorithm [10] whose only drawback is its low speed: it is an $O(N^2)$ algorithm. Nevertheless, the algorithm is well suited for research purposes.

The algorithm solves the following problem:

Given are two sets M^+ and M^- of points in an n-dimensional space.

Find an $(n-1)$-dimensional hyperplane HP separating the sets.

The solution: HP is specified by two vectors A^+ and A^- as the middle perpendicular to the line segment (A^+, A^-). Let $Dist(P)$ be the signed distance of the point P to HP.

Figure 7.14 shows an example of two sets of points in a 2D space.

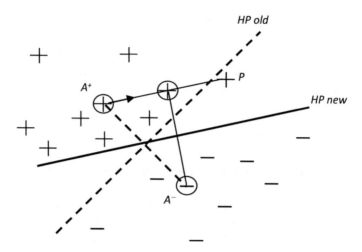

Fig. 7.14 An example of the correction of the separating plane

The Algorithm:

1. Set A^+ equal to an arbitrary point from M^+ and A^- equal to an arbitrary point from M^-. Carry out a sequence of the following iterations.
2. During each iteration test all points P from M^+ and M^- as follows:

    If the point $P \in M^+$ lies on the wrong side of HP which means $Dist(P) < 0$, then set:

    $A^+=$ *Foot of the perpendicular from* A^- *to the segment* (A^+, P).

    If $P \in M^-$ and $Dist(P) > 0$, then set:

    $A^-=$ *Foot of the perpendicular from* A^+ *to the segment* (A^-, P).

3. If there is no point on the wrong side of HP stop the Algorithm. The separating hyperplane is the middle perpendicular to the line segment (A^+, A^-).
4. If the distance between A^+ and A^- is less than a predefined threshold, then there exists no separating hyperplane; the convex hulls of M^+ and M^- intersect.

Figure 7.14 shows an example of separating two sets of points in a 2D space. The sets are represented by "+" and "−" signs, the vectors A^+ and A^- are represented by encircled signs; the old and the new separation plane are presented by a dashed and a solid line. The point P lies on the wrong side of the old plane. It is connected with the old vector A^+ and a perpendicular has been dropped from A^- onto the segment (A^+, P). The foot of the perpendicular is the new vector A^+.

We have applied this method to recognize DPPs while using the set of the positive voxels V^+ as M^+ and that of the negative voxels V^- as M^-.

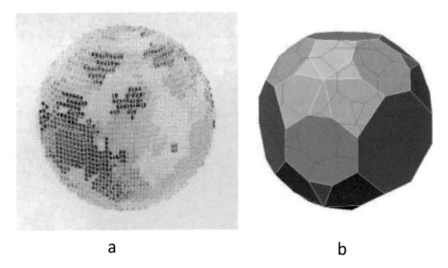

a   b

Fig. 7.15 Examples of segmenting the boundary of a digital ball into DPPs by arbitrarily choosing the next facet (**a**) and by computing its convex hull (**b**)

7.3.4 The Partial Problem of the "Choice"

There is no efficient method for the solution of the partial problem "Choice" known until now. It should be mentioned that this problem is much more difficult in the case of DPPs as in the case of DSSs. Really, in the latter case there are exactly two possibilities to continue a partially recognized DSS: forward or backward along the digital curve. But in the case of a DPP there are as many possibilities to continue as the number of facets adjacent to a partially recognized DPP. There is no known criterion to decide which of them should be preferred. When choosing the next facet arbitrarily, then the found DPPs look chaotic even for surfaces of regular polyhedrons (Fig. 7.15).

References

1. Rosenfeld A, Kong Y, Wu A. Digital Surfaces. CVGIP, Graphical Models. 1991;53:305–12.
2. Gordon D, Udupa JK. Fast surface tracking in three-dimensional binary images. CVGIP. 1989;45:196–214.
3. Aho AV, Hopcroft JH, Ullman JD. The design and analysis of algorithms. Reading: Addison-Wesley; 1974.
4. Kovalevsky V. Geometry of locally finite spaces. Editing house Dr. Bärbel Kovalevski, Berlin; 2008. ISBN 978-3-9812252-0-4
5. Kong TY. On boundaries and boundary crack codes in multidimensional digital images. In: Shape in picture. Springer; 1994. p. 71–80.

6. Urbanek C. Visualizing of the Kovalevsky's algorithm for surface presentation. Master thesis: University of Applied Sciences, Berlin; 2003.
7. Kovalevsky V. A topological method of surface representation. In: Bertrand G, Couprie M, Perroton L, editors. DGCI, LNSC 1568; 1999. p. 118–35.
8. Klette R, Sun HJ. Digital planar segment based Polyhedrization for surface area estimation. In: Proc. visual form 2001. LNCS 2059; 2001. p. 356–66.
9. Reveilles JP. Géométrie discrète, calcul en nombres entier et algorithmique. Thèse d'état: Université Louis Pasteur, Strasbourg; 1991.
10. Kozinets BN. An iteration-algorithm for separating the convex hulls of two sets. In: Vapnik VN, editor. Lerning algorithms for pattern recognition (in Russian language). Moscow: Publishing house "Sovetskoe Radio"; 1973.

Chapter 8
Edge Detection in 3D Images

Abstract This chapter describes a new algorithm for the edge detection in 3D images. As in the 2D case, edge detection in 3D images needs a preprocessing of the image: a 3D version of the Sigma filter to reduce the Gaussian noise and the 3D version of the Extreme filter to sharpen the edges. Algorithms for the 3D Sigma filter, the 3D Extreme filter, and for the 3D edge detection are described.

Keywords Edge detection in 3D · Preprocessing of 3D images · Sigma3D · Extrem3D · Algorithm Edge3D

We have described in Sect. 6.2 a new method of edge detection. This method can be used also for the edge detection in 3D images. It is simpler than the edge detection in 2D images because a 3D image does not contain colors, but only densities.

As in the 2D case, edge detection demands certain preprocessing of the image: it is necessary to suppress the Gaussian noise and to sharpen the ramps. Algorithms like the Sigma-Filter and the Extreme Filter used in the 2D case are described in the next section.

8.1 Preprocessing of 3D Images

We present here the algorithm of the 3D Sigma filter. The algorithm is like the 2D algorithm "SigmaGray" described in Sect. 6.1.1. The 3D algorithm differs from the 2D one by two additional repetitions: Repetition with the coordinate "Z" of the central voxel of the gliding three-dimensional window and the repetition with the coordinate "z" of the voxel gliding through the window. The fields "Image" and "Output" are three-dimensional fields with the sizes Width, Height, and Depth. The field "Output" is global.

8.1.1   Algorithm Sigma3D

Algorithm Sigma3D(Field Image[Width, Height, Depth], integer hWind, integer Tol)

Step 1: Define the variables CenterV, MaxDif, MinDif, xEnd,
       xStart, yEnd, yStart, zEnd, zStart, Sum, NumSum, x, X, y, Y, z, and Z.
Step 2: Set Z ← 0,
Step 3: Repeat until Z is less than Depth:
    3.1: Set zStart ← Max(0, Z – hWind), EndJ ← Min(Depth – 1, Z + hWind),
        and Y ← 0.
    3.2: Repeat until Y is less than Height:
    3.2.1: Set yStart ← Max(0, Y – hWind),
        yEnd ← Min(Height – 1, Y + hWind), and X ← 0.
    3.2.2: Repeat until X is less than Width:
    Begin 3.2.2
       3.2.2.1: Set CenterV ← Image[X, Y, Z], Sum ← 0, NumSum ← 0,
        MinDif ← Max(0,CenterV–Tol), and MaxDif ← Min(255,CenterV+Tol).
       3.2.2.2: Set z ← zStart.
       3.2.2.3: Repeat until z is less or equal to zEnd:
       3.2.2.3.1: Set y ← yStart.
       3.2.2.3.2: Repeat until y is less or equal to yEnd:
        3.2.2.3.2.1: Set x ← xStart.
        3.2.2.3.2.2: Repeat until x is less or equal to xEnd:
        Begin
        If Image[x, y, z] is between MinDif and MaxDif, then
          set Sum ← Sum + Image[x, y, z] and NumSum ← NumSum + 1.
        Set x ← x + 1.
        End   (End of Repeat 3.2.2.3.4)
       3.2.2.3.2.3: Set y ← y + 1.   (End of Repeat 3.2.2.3.2)
       3.2.2.3.3: Set z ← z + 1.   (End of Repeat 3.2.2.3)
       3.2.2.4: If NumSum is positive, then set Output[X, Y, Z] ← Sum/NumSum,
        else set Output[X, Y, Z] ← Image[X, Y, Z].
       3.2.2.5: Set X ← X + 1.   (End of Repeat 3.2.2)
    End 3.2.2
    3.2.3: Set Y ← Y + 1.   (End of Repeat 3.2)
    3.3: Set Z ← Z + 1.   (End of Repeat 3)
    3.2:
Stop

Now we present the algorithm of the extreme filter for 3D images. The algorithm "Extrem3D" is like the 2D algorithm "ExtremGray" described in Sect. 6.1.2. The 3D algorithm differs from the 2D one by two additional repetitions: Repetition with the coordinate "Z" of the central voxel of the gliding three-dimensional window and the repetition with the coordinat "z" of the voxel gliding through the window.

8.1.2   *Algorithm Extrem3D*

Algorithm Extrem3D(Image[width, height, depth], int hWind)

Step 1: Define the variables Dens, DensMax, DensMin, x, xL, xR, X, y, Y, yEnd, yStart, z, Z, zEnd and zStart.

Step 2: Define the field Histo[256].

Step 3: Set Z ← 0.

Step 4: Repeat until Z is less than Image.depth:

    4.1: Set zStart ← Max(0, Z – hWind) and zEnd ← Min(Image.height, Z + hWind).

    4.2: Set Y ← 0.

    4.3: Repeat until Y is less than Image.height:

      4.3.1:Set yStart ← Max(0, Y – hWind) and yEnd ← Min(Image.height, Y + hWind).

      4.3.2: Set X ← 0.

      4.3.3: Repeat until X is less than Image.width:

    :4.3.3.1:If X is equal to 0, then

        Begin 4.3.3.1

        Set all element of Histo equal to 0.

        Set z ← zStart.

        Repeat until z is less or equal to zEnd:

          Begin z

          Set y ← yStart.

          Repeat until y is less or equal to yEnd:

            Begin y

            Set x ← 0.

            Repeat until x is less or equal to hWind:

              Begin x

              Set Histo[Image[x, y, z]] ← Histo[Image[x, y, z]] + 1.

              Set x ← x + 1.

              End x   (End of Repeat until x is less ...)

              Set y ← y + 1.

            End y   (End of Repeat until y is less ...)

            Set z ← z + 1.

          End z   (End of Repeat until z is less ...)

        End 4.3.3.1   (End step 4.3.3.1; If X is equal to 0)

    4.3.3.2: If X is greater than 0, then

      Begin 4.3.3.2

      Set xR ← X + hWind.

      If xR is less than Image.width, then

      Begin If xR

      Set z ← zStart.

      Repeat until z is less than or equal to zEnd:

        Begin z

        Set y ← yStart.

        Repeat until y is less than or equal to yEnd:

```
            Begin y
              Set Histo[Image[x,y,z]] ← Histo[Image[x,y,z] ]+ 1. (Increasing right plane)
              Set y ← y + 1.
            End y              (End repeat until y is less …)
            Set z ← z + 1.
         End z                 (End of Repeat until z is less …)
      End If xR                (End of If xR is less than Image.width)

   Set xL ← X – hWind - 1.
   If xL is greater or equal to 0, then
     Begin If xL
       Set z ← zStart.
       Repeat until z is less than or equal to zEnd:
         Begin z
           Set y ← yStart.
           Repeat until y is less than or equal to yEnd:
             Begin y
               Set Histo[Image[x,y,z]] ← Histo[Image[x,y,z]] - 1. (Decreasing right plane)
               Set y ← y + 1.
             End y             (End repeat until y is less …)
             Set z ← z + 1.
           End z               (End of Repeat until z is less …)
         End If xL             (End of If xL is greater or equal to 0)
       End 4.3.3.2             (End step 4.3.3.2: If X is greater than 0)

   4.3.3.3:  Set Dens ← 0.
     Repeat until Dens is less than 256:
       Begin
         If Histo[Dens] is positive, then set DensMin ← Dens and interrupt repetition.
         Set Dens ← Dens + 1.
       End                     (End of Repeat until Dens is less than 256)

   4.3.3.4: Set Dens ← 255.
     Repeat until Dens is greater or equal to 0:
       Begin
         If Histo[Dens] is positive set DensMax ← Dens and interrupt repetition.
         Set Dens ← Dens - 1.
       End                       (End of Repeat until Dens is greater or equal to 0)
   4.3.3.5: If Image[X, Y, Z] – DensMin is less than DensMax - Image[X, Y, Z], then
                   Set Output[X, Y, Z] ← DensMin,
                else set Output[X, Y, Z] ← DensMax.
   4.3.3.6: Set X ← X + 1.                      (End of repeat Step 4.3.3)
   4.3.4:  Set Y ← Y + 1.                        (End of repetitio Step 4.3)
   4.4: Set Z ← Z + 1.                           (End of repeat Step 4)
Stop
```

We present below the algorithm for the 3D edge detection. The algorithm "Edge3D" is like the 2D algorith "EdgeDetect" described in Sect. 6.2. The 3D algorithm differs from the 2D one by the great additional group of repetitions with the steps 11, 11.2, and 11.2.2 going through facets orthogonal to the coordinate axis Z. The two-dimensional algorithm contains only repetitions through cracks orthogonal to the coordinate axes X and Y.

The edges are represented as labeled facets and cracks in the cell complex "Complex3D". The utility algorithm "ToEdge3D" is like to the 2D version presented above in Sect. 6.2.

8.2 Algorithm Edge3D

Algorithm Edge3D(Image[width, height, depth], Threshold, CNX, CNY, CNZ):

Step 1: Define the variables Code, DensP, DensV, i, x, y, z, x1, y1, and z1.
Step 2: Set Code ← 0.
Step 3: Define the fields DIF[5] and ADIF[5].
Step 4: Define and initialize constant field Bit[6] ← {0, 1, 2, 4, 8, 16}.
Step 5: Define and initialize constant field LUT:
    LUT ← { 0, 0, 0, 0, 0, 2, 2, 3, 5,
           0, 0, 0, 0, 1, 1, 4, 9,
           0, 0, 0, 0, 2, 2, 3, 5,
           0, 0, 0, 0, 6, 6, 7, 8};
Step 6: Set z ← 1.
Step 7: Repeat until z is less than CNZ:
 Step 7.1: Set y ← 1.
 Step 7.2: Repeat until y is less than CNY:
  Step 7.2.1: Set x ← -2.
  Step 7.2.2: Repeat until x is less than CNX - 1:   (Though X-facets (x, y, z)
   Begin 7.2.2.1
    Set x1 ← x + 4.
    If x1 is less than CNX − 1, then
    Begin
     Set DensP ← Image[(x1+1)/2, y/2, z/2].
     Set DensV ← Image[(x1-1)/2, y/2, z/2].
     Set DIF[0] ← DensV − DensP.
     Set ADIF[0] ← absolute value of DIF[0].
    End
    else DIF[0] ← 0 and ADIF[0] ← 0.
    Set Code ← 0.
    Set i ← 0.
    Repeat until i is less than 5:
     Begin
      If ADIF[i] is greater than Threshold, then set Code ← Code + Bit[i+1].
      Set i ← i 1.
     End   (End Repeat until i is less than 5)
    If the function ToEdge(LUT[Code], ADIF, DIF) returns true, then:
     Begin
      Set Complex3D[x, y, z] ← 1.   (Labeling the facet (x, y, z) as edge facet)
      Set Complex3D[x, y+1, z] ← Complex3D[x, y+1, z] + 1.   (The XY crack)
      Set Complex3D[x, y-1, z] ← Complex3D[x, y+1, z] + 1.   (The XY crack)
      Set Complex3D[x, y, z+1] ← Complex3D[x, y+1, z] + 1.   (The XZ crack)
      Set Complex3D[x, y, z-1] ← Complex3D[x, y+1, z] + 1.   (The XZ crack)
     End
    Set i ← 4.
    Repeat until i remains greater than 0:

Begin
   ADIF[i] ← ADIF[I − 1].
   DIF[i] ← DIF[I − 1].
   Set i ← i − 1.
End   (End Repeat until i remains greater than 0)
  Set x ← x +2.
 End 7.2.2.1. (End repetition 7.2.2)
 Step 7.2.3: Set y ← y +2. (End repetition 7.2)
Step 7.3: Set z ← z +2. (End repetition 7)

Step 8: Set z ← 1.
Step 9: Repeat until z is less than CNZ:
 Step 9.1: Set x ← 1.
 Step 9.2: Repeat until x is less than CNX:
  Step 9.2.1: Set y ← -2.
  Step 9.2.2: Repeat until y is less than CNY - 1:   (Though Y-facets (x, y, z)
  Begin 9.2.2
   Set y1 ← y + 4.
   If y1 is less than CNY − 1, then
    Begin
     Set DensP ← Image[x/2, (y1 + 1)/2, z/2].
     Set DensV ← Image[x/2, (y1 − 1)/2, z/2].
     Set DIF[0] ← DensV − DensP.
     Set ADIF[0] ← absolute value of DIF[0].
    End
   else DIF[0] ← 0 and ADIF[0] ← 0.
   Set Code ← 0.
   Set i ← 0.
   Repeat until i is less than 5:
    Begin
     If ADIF[i] is greater than Threshold, then set Code ← Code + Bit[i+1].
     Set i ← i + 1.
    End   (End Repeat until i is less than 5)
   If the function ToEdge(LUT[Code], ADIF, DIF) returns true, then:
    Begin
     Set Complex3D[x, y, z] ← 1. (Labeling the facet (x, y, z) as edge facet)
     Set Complex3D[x+1, y, z] ← Complex3D[x, y+1, z] + 1. (The YX crack)
     Set Complex3D[x-1, y, z] ← Complex3D[x, y+1, z] + 1. (The YX crack)
     Set Complex3D[x, y, z+1] ← Complex3D[x, y+1, z] + 1. (The YZ crack)
     Set Complex3D[x, y, z-1] ← Complex3D[x, y+1, z] + 1. (The YZ crack)
    End
   Set i ← 4.
   Repeat until i remains greater than 0:
    Begin
     ADIF[i] ← ADIF[i − 1].
     DIF[i] ← DIF[i − 1].
     Set i ← i - 1.
    End   (End Repeat until i remains greater than 0)

Set y ← y +2.
   End 9.2.2   (End repetition 9.2.2)
   Step 9.2.3: Set x ← x +2.   (End repetition 9.2)
   Step 9.3: Set z ← z +2.   (End repetition 9)

Step 10: Set y ← 1.
Step 11: Repeat until y is less than CNY:
 Step 11.1: Set x ← 1.
 Step 11.2: Repeat until x is less than CNX:
  Step 11.2.1: Set z ← -2.
  Step 11.2.2: Repeat until z is less than CNZ - 1:   (Though Z-facets (x, y, z)
   Begin 11.2.2
    Set z1 ← z + 4.
    If z1 is less than CNZ – 1, then
    Begin
     Set DensP ← Image[x/2, y/2, (z1 + 1)/2].
     Set DensV ← Image[x/2, y/2, (z1 – 1)/2].
     Set DIF[0] ← DensV – DensP.
     Set ADIF[0] ← absolute value of DIF[0].
    End
    else DIF[0] ← 0 and ADIF[0] ← 0.
    Set Code ← 0.
    Set i ← 0.
    Repeat until i is less than 5:
     Begin
      If ADIF[i] is greater than Threshold, then set Code ← Code + Bit[i+1].
      Set i ← i + 1.
     End
    If the function ToEdge(LUT[Code], ADIF, DIF) returns true, then:
     Begin
      Set Complex3D[x, y, z] ← 1.   (Labeling the facet (x, y, z) as edge facet)
      Set Complex3D[x+1, y, z] ← Complex3D[x, y+1, z] + 1. (The ZX crack)
      Set Complex3D[x-1, y, z] ← Complex3D[x, y+1, z] + 1. (The ZX crack)
      Set Complex3D[x, y+1, z] ← Complex3D[x, y+1, z] + 1. (The ZY crack)
      Set Complex3D[x, y-1, z] ← Complex3D[x, y+1, z] + 1. (The ZY crack)
     End
    Set i ← 4.
    Repeat until i remains greater than 0:
     Begin
      ADIF[i] ← ADIF[i – 1].
      DIF[i] ← DIF[i – 1].
      Set i ← i - 1.
     End
    Set z ← z +2.
   End 11.2.2   (End repetition 11.2.2)
  Step 11.2.3: Set x ← x +2.   (End repetition 11.2)
 Step 11.3: Set y ← y +2.   (End repetition 11)
Stop

We have prepared an artificial 3D image of the size of 100*100*100 voxels
containing a parallelepiped of the size 50*50*50 voxels rotated by 45° about an
axis parallel to the Z axis. The image contains additionally an adjacent cylinder with
a horizontal axis perpendicular to one side of the parallelopiped and going through

its central point. Figure 8.1 shows the intersection of the edge of the body with the diagonal plane of the space.

Figure 8.2 shows the intersection of the edge of the cylinder with a plane orthogonal to its axis.

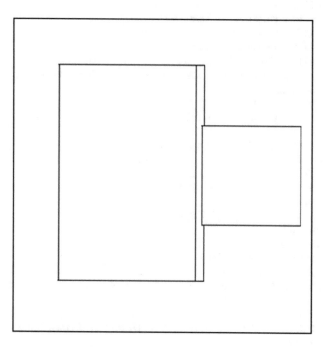

Fig. 8.1 Intersection of the edge of the body with the diagonal plane of the space

Fig. 8.2 Intersection of the edge of the cylinder with the plane orthogonal to its axis

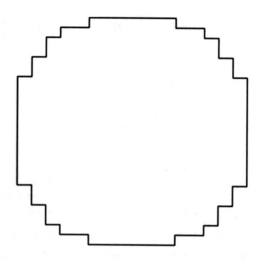

Chapter 9
Discussion

Abstract This chapter explains that the estimation of derivatives by means of the classical definition of a derivative as the limit of the relation of the increment of the function to the increment of the argument while the latter increment tends to zero, can lead to errors. Suggested is to use the relation of the increment of the function to the *optimal* increment of the argument, rather than to a small increment. A method of calculating the optimal increment of the argument is suggested.

Keywords Estimation of derivatives · Limit at increment tending to zero · Errors at small increments · Optimal increment · Calculation of the optimal increment

9.1 Optimal Increment for Calculating Derivatives

Consider the problem of estimating derivatives of non analytical functions. The classical definition considers the derivative as the limit of the expression:

$$(f(x + \Delta x) - f(x))/\Delta x \tag{9.1}$$

as Δx tends to zero. This definition is based on the supposition that the value of the function can be calculated with an arbitrarily high precision. To see that this supposition is important, consider an experiment of numerically calculating by a computer the value of the first derivative of the function $y = f(x) = x^2$ at $x = 1.5$, which one knows to be equal to 3.0. The computer must calculate for this purpose the value of (9.1) for $x = 1.5$ and for as little as possible values of Δx. Figure 9.1 presents the results of computing this value in the interval [1.3, 1.7] with the values of $\Delta x = 10^{-p}$, with $p = 16, 17, 18$, and 19. As one can see in Fig. 9.1 below, at first the results become inaccurate and then they become crazy. The reason for the errors is that at values of Δx comparable with the rounding error of $f(x)$ the result of the division becomes first imprecise and then meaningless.

The following is a citation from [1].

To investigate this phenomenon let us consider the influence of errors in representing $f(x)$ onto the value of (9.1):

© The Author(s), under exclusive license to Springer Nature Singapore Pte Ltd. 2021
V. Kovalevsky, *Image Processing with Cellular Topology*,
https://doi.org/10.1007/978-981-16-5772-6_9

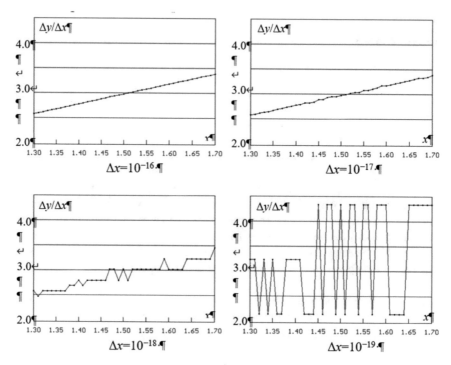

Fig. 9.1 Estimating the first derivative of $y = x^2$ as $\Delta y/\Delta x$ with small values of Δx

$$\text{Estimate}(f'(x)) = (f(x + \Delta x) + er_1 - f(x) - er_2)/\Delta x$$
$$= (f(x + \Delta x) - f(x))/\Delta x + (er_1 - er_2)/\Delta x; \qquad (9.2)$$

where er_1 and er_2 are the errors in computing $f(x + \Delta x)$ and $f(x)$ respectively.

The Table 9.1 shows the calculated values "Der" of the derivative and its deviation "Der – 3" from the true value 3 for different values "h" of the increment Δx in (9.1).

It is possible to find the optimum value of Δx depending on the imprecision of the values of the function and on the values of its higher differences. If these errors are limited by the value of ε, then the worst-case error E of the estimate is equal to

$$Er = 2 \cdot \varepsilon/\Delta x. \qquad (9.3)$$

As soon as Δx becomes of the order of magnitude of ε, the error Er of the estimate becomes unacceptably great.

According to (9.3) the error Er may be made arbitrarily small by increasing Δx. However, there is another source of errors, which must be considered: If the desired derivative is not constant (i.e., $f(x)$ is not a linear function), then the estimate (9.1) brings the *average* derivative in the interval $[x, x + \Delta x]$, which may be essentially different form the value of the derivative at the point x. It is known that the estimate

Table 9.1   Values of the derivative "Der" and its deviation Der - 3 from the true value

```
h=0,1; Der=3,10000000000001; Der - 3 =0,100000000000005
h=0,01; Der=3,01; Der - 3 =0,0100000000000016
h=0,001; Der=3,00099999999981; Der - 3 =0,000999999999809376
h=0,0001; Der=3,00010000000128; Der - 3 =0,000100000001282297
h=1E-05; Der=3,00001000002048; Der - 3 =1,00000204805717E-05
h=1E-06; Der=3,00000099962006; Der - 3 =9,99620056063577E-07
h=1E-07; Der=3,00000010167167; Der - 3 =1,0167167374675E-07
h=1E-08; Der=2,99999998176759; Der - 3 =-1,82324129127664E-08
h=1E-09; Der=3,00000024822111; Der - 3 =2,48221112997271E-07
h=1E-10; Der=3,00000024822111; Der - 3 =2,48221112997271E-07
h=1E-11; Der=3,00000024822111; Der - 3 =2,48221112997271E-07
h=1E-12; Der=3,00026670174702; Der - 3 =0,000266701747023035
h=1E-13; Der=2,99760216648792; Der - 3 =-0,00239783351207734
h=1E-14; Der=3,01980662698043; Der - 3 =0,0198066269804258
h=1E-15; Der=3,5527136788005; Der - 3 =0,5527136788005
h=9E-16; Der=2,96059473233375; Der - 3 =-0,0394052676662495
h=8E-16; Der=3,33066907387547; Der - 3 =0,330669073875469
h=7E-16; Der=3,1720657846433; Der - 3 =0,172065784643304
h=6E-16; Der=3,70074341541719; Der - 3 =0,700743415417188
h=5E-16; Der=2,66453525910038; Der - 3 =-0,335464740899624
h=4E-16; Der=3,33066907387547; Der - 3 =0,330669073875469
h=3E-16; Der=2,96059473233375; Der - 3 =-0,0394052676662504
h=2E-16; Der=4,44089209850062; Der - 3 =1,44089209850062
h=1E-16; Der=0; Der - 3 =-3
h=0; Der=n. def.; Der - 3 =n. def.
h=-9,99999999999999E-17; Der=0; Der - 3 =-3
h=-2E-16; Der=2,22044604925031; Der - 3 =-0,779553950749686
h=-3E-16; Der=1,48029736616688; Der - 3 =-1,51970263383312
h=-4E-16; Der=3,33066907387547; Der - 3 =0,330669073875471
h=-5E-16; Der=2,66453525910038; Der - 3 =-0,335464740899623
```

OK

(9.1) is equal to the exact value of the derivative at some unknown point in the said interval. When increasing Δx we increase the precision of the desired estimate, however, we simultaneously decrease the precision of the *location* of the estimate. Thus Eq. (9.3) represents some kind of *principle of uncertainty* (which may have much in common with the famous Heisenberg uncertainty principle in the quantum mechanics): the greater the precision of the location of the estimate, the smaller the precision of the estimate itself.

It is easy to see that there is an *optimum value* of Δx yielding the highest precision. We consider here the *maximum possible error* of the estimate rather than an RMS (root mean square) error because knowing the maximum error gives us a *guaranty* that the actual error will never override the found limit.

To find the optimum value of Δx consider the worst-case error of the estimate (9.2) while representing $f(x)$ by its Taylor series up to the second order term:

$$\text{Estimate } (f'(x)) = (f(x + \Delta x) + er_1 - f(x) - er_2)/\Delta x =$$
$$= f'(x) + 0.5 \cdot f''(x + k \cdot \Delta x) \cdot \Delta x + (er_1 - er_2)/\Delta x;$$

where k is a value between 0 and 1.

The error *Er* of the estimate is then equal to:

$$Er = \text{Estimate } (f'(x)) - f'(x) = 0.5 \cdot f''(x + k \cdot \Delta x) \cdot \Delta x + (er_1 - er_2)/\Delta x.$$

To get the maximum absolute error we should substitute the maximum value *F2* of $f''(x)$ in the interval $[x, x + \Delta x]$ for $f''(x + k \Delta x)$, ε for er_1 and $-\varepsilon$ for er_2:

$$maxEr = 0.5 \cdot F2 \cdot \Delta x + 2 \cdot \varepsilon/\Delta x. \tag{9.4}$$

To get the optimal value of Δx we must find the minimum of (9.4) with respect to Δx, i.e., to set the partial derivative of (9.4) with respect to Δx equal to 0 and to solve the corresponding equation:

$$\partial maxEr/\partial \Delta x = 0.5 \cdot F2 - 2 \cdot \varepsilon/\Delta x^2 = 0.$$

It follows:

$$\text{optimal } \Delta x = 2 \cdot \sqrt{\varepsilon/F2}; \tag{9.5}$$

where ε is the maximum possible error of specifying the values of $f(x)$ and *F2* is the maximum possible value of the second derivative of $f(x)$ in the interval $[x, x + \Delta x]$. Since we are mostly interested to know and to minimize the *order of magnitude* of the maximum error (9.5) rather than its exact value, the parameters ε and *F2* in the right-hand side of (9.6) may be estimated rather coarsely. The value of *F2* can be estimated as

$$\text{Estimate of } F2 = (f(x + \Delta x) - 2 \cdot f(x + \Delta x) + f(x - \Delta x))/\Delta x^2. \tag{9.6}$$

To illustrate the usage of (9.6) let us apply it to the case of Fig. 9.1. When computing $f(x)$ as a double precision float with a mantissa of 53 bits, then the value ε is equal to the rounding error and is of the order of $2^2 2^{-54} = 2^{-52} \cong 10^{-16}$. Note that this is not the smallest number representable in the computer, which is much less. The smallest number is specified by the exponent, rather than by the mantissa.

The optimal value of Δx is, according to (9.6) of the order of 10^{-8}. The error in computing the derivative at $x = 1.5$ becomes unacceptably great when Δx becomes comparable with or smaller than $2\varepsilon = 2 \cdot 10^{-16}$ (since the second term in (9.4) becomes greater than 1.0), which situation we see in Fig. 9.1.

The error may be made still smaller if we use a symmetrical estimate of the first derivative:

$$\text{Estimate1S} = (f(x + \Delta x) - f(x - \Delta x))/(2 \cdot \Delta x); \qquad (9.7)$$

The optimal value of Δx may be deduced in the same way as before; however, the Taylor series should be prolonged until the term with the third derivative since the terms with the second derivative disappear. Then we obtain:

$$\text{Error} = (1/6) \cdot F3 \cdot \Delta x^2 + \varepsilon/\Delta x; \qquad (9.8)$$

and

$$\text{optim } \Delta x = (3 \cdot \varepsilon/F3)^{1/3}; \qquad (9.9)$$

where $F3$ is the estimate of the maximum absolute value of the third derivative of $f(x)$ in the interval $[x - \Delta x, x + \Delta x]$. The estimate of $F3$ can be found in a similar way as the estimate of $F2$ according to (9.7).

It is important to notice that in the case $F3 = 0$ the value of Δx should be chosen as great as possible. To make this clear, the reader may persuade himself, that the estimate (9.7) yields the *precise value* of the first derivative at the point x with any value of Δx if $f(x)$ is a polynomial of degree 2.

Now we can specify the worst-case error when using the optimal value of Δx. We put the value (9.9) into (9.8) and obtain:

$$\text{minimax Error} = ((1/6) \cdot 3^{2/3} + 3^{-1/3}) \cdot \varepsilon^{2/3} \cdot F3^{1/3}$$
$$\cong 1.04 \cdot \varepsilon^{2/3} \cdot F3^{1/3}. \qquad (9.10)$$

The inference of the optimal values of Δx for the second derivative can be found in [1].

One may conclude from the above considerations that in cases when the values of a function can be computed with limited precision, the computation of its derivatives must be performed according to the above expression (9.1) or similar expressions for higher derivatives with the optimal value of Δx, whereas the classical derivatives may serve only as *approximations* which are the more precise, the higher the precision of the values of the function and the lower the absolute values of higher derivatives. (It is easily recognized e.g. that the precision of calculating by means of (9.1) the first derivative of a linear function, whose higher derivatives are equal to zero, is the higher the *greater* the absolute value of Δx).

9.2   Conclusion

The estimates of derivatives should be performed not as a limit of expressions like (9.1) with Δx tending to zero but rather by calculating the value of (9.1) with the optimal value of Δx as explained above.

Reference

1. Kovalevsky V. Curvature in digital 2D images. Int J Pattern Recognit Artif Intell. 2001;15(7): 1183–200.

Printed in the United States
by Baker & Taylor Publisher Services